The Advance of Neuroscience

The Advance of Neuroscience

Twelve Topics from the Victorian Era to Today

Lori A. Schmied

McFarland & Company, Inc., Publishers
Jefferson, North Carolina

Library of Congress Cataloguing-in-Publication Data

Names: Schmied, Lori A., 1958– author.
Title: The advance of neuroscience : twelve topics from the Victorian era to today / Lori A. Schmied.
Description: Jefferson, North Carolina : McFarland & Company, Inc., 2019 | Includes bibliographical references and index.
Identifiers: LCCN 2018049361 | ISBN 9781476665573 (softcover : acid free paper) ♾
Subjects: | MESH: Neurosciences—history | History, 19th Century | History, 20th Century | History, 21st Century | Brain | Psychotropic Drugs—adverse effects | United States
Classification: LCC QP360 | NLM WL 11 AA1 | DDC 612.8/233—dc23
LC record available at https://lccn.loc.gov/2018049361

British Library cataloguing data are available

ISBN (print) 978-1-4766-6557-3
ISBN (ebook) 978-1-4766-3397-8

Printed in the United States of America

McFarland & Company, Inc., Publishers
Box 611, Jefferson, North Carolina 28640
www.mcfarlandpub.com

Table of Contents

Acknowledgments

MANY INDIVIDUALS HAVE HELPED ME to bring this work to fruition and deserve my gratitude. I am especially grateful to my friends and colleagues in psychology at Maryville College for the support they have given me not just for this project but throughout my academic career. I have learned much about the field of psychology from them and they have helped me become a better teacher. The friendship and support of my chair, Kathie Shiba, has been particularly invaluable in completing this project. My colleagues at the Lamar Memorial Library at Maryville fulfilled countless interlibrary loan requests and helped track down sources.

I would also like to thank the staff at the Wellcome Library in London for their assistance. I have spent many hours during the summer months over the last decade combing through resources at this library dedicated to the history of medicine. Its vast holdings enabled me to conduct most of my research under one roof and its digital holdings allowed access when I was at home.

The annual meetings of the Southern Society for Philosophy and Psychology were a means to develop many of the topics that make up some of the chapters in this book. Many, if not most, of the chapters were initially presented in abbreviated form to members of that society and their feedback was most helpful. They were also supportive of my efforts as a "newbie" to the history of psychology and helped give me confidence that I could pursue a project of this scale.

My psychology and neuroscience students past and present at Maryville also deserve my thanks. My first time teaching History and Systems of Psychology was transformative in sparking my interest in the history of psychology and providing a new direction of research. I continue to be inspired by my students and challenged to think more deeply on the changing field of neuroscience. Special thanks go to my former student Katherine McNeely White, who is now a graduate student in cognitive neuroscience, for her assistance in locating contemporary sources.

Our trips to London for research over the years were made possible in part because of our friendship with Camilla Bosanquet and her family, who first opened their home to our family 35 years ago when we came for a sabbatical year for my husband. Without periodic support by the Maryville College Faculty Development Committee to fund summer research trips to the Wellcome Library, this project would have stalled long ago. My early mentors, Kathleen Lawler, Hannah Steinberg, and Jerry Waters, each in their own way taught me to value the history of psychology and to search for connections at the intersections of disciplines.

Finally, my deepest gratitude goes to my husband and daughter, Karl and Katie, who have been my biggest cheerleaders throughout this process. Their support has been unflagging, giving me encouragement at the slightest hint of faltering. Karl has been my sounding board and advisor whenever I needed a boost or an idea to break through a research dead-end. No task has been too small or too big to offer help. Their love and support have made me a better person.

Preface

ON ANY GIVEN DAY, YOU CAN BE SURE to find at least one popular media headline about some topic related to neuroscience. Just today, I have found articles in major newspapers on dementia, Alzheimer's, the opiate addiction crisis, and the increase in prescription drugs to treat ADHD. The public has a fascination with brain science, and rightly so—the advances in our knowledge of the inner workings of the brain have been substantial in the last 20 years, surpassing the accumulated knowledge of the last several centuries! One only has to use the prefix "neuro-" and add a picture of a brain image to have the news media and public's attention. The appetite for brain news is voracious—the public, it seems, cannot get enough of brain matter. The same can be said for academia. New areas of study, all with "neuro" prefixes, abound: neuroeconomics, neuroaesthetics, neurohistory, neuroforensics. We are even now in the midst of an immensely ambitious ten-year federal initiative created by President Barack Obama to map the brain. A few commentators have referred to these developments as the "neuro-turn," and everywhere one looks, it is now with a "neuromolecular gaze."[1]

Some discoveries mean discarding old theories while others resurrect ones long forgotten. A look at the history of neuroscience reveals common themes, perennial debates, and a continuing quest to understand the relationship of body and mind. This book explores some of those themes and shows how history can bring contemporary issues into sharper relief. Neuroscience is a field rife with potential for both elucidation and exploitation. Instead of uncritical acceptance of new findings and applications, a look into the history of neuroscience can help one to sort through the bewildering array of issues.

The advances in science and technology have been so substantial over the last few decades that there is a tendency to view the present day as the natural culmination of steps in a journey to some predetermined final destination. This naïve view ignores the history of scientific thought and the many false starts and dead-ends. The field of neuroscience may be an especially good example. The last two decades have seen an explosion in our knowledge of brain function using a variety of scientific disciplines.

I felt compelled to take on this work not just because of my professional research in the history of neuroscience issues, particularly psychopharmacology, but from my 30 years of experience teaching in the field. There is a need for educators to make neuroscience accessible to an audience that extends beyond one's colleagues in a narrow niche of academia. The very interdisciplinary nature of neuroscience lends itself to a broader audience and an historical account helps construct those disparate connections among the various fields of study that are contemporary neuroscience into a coherent package. My own background in biopsychology, psychopharmacology, and psychophysiology has enhanced my understanding of some of these connections. In my role of teacher, I am asked on a regular basis by students, colleagues, and acquaintances alike to explain a neuroscience concept or comment on some development or research advance reported in the news. While the media and public may crave simple answers to complex problems, that is an unrealistic expectation in neuroscience. Yet we need to be able to translate these complicated issues into comprehensible terms for non-experts. I believe that educators are uniquely positioned to convert the convoluted and sometimes opaque technical language of a complex research issue into a decipherable account that does justice to the problem at hand. The overlay of an historical account provides a thread of continuity—one quickly sees that seemingly disparate issues are in fact related by similar roots in historical assumptions about the nervous system, for example. Neuro-turn aficionados from the humanities and arts may have the zealous and uncritical enthusiasms of the newly converted and sometimes overstate the promise of neuroscience, while the neuroscientists themselves sometimes lack the background to apply historical and socio-cultural context to an issue in order to understand its limitations and embedded assumptions.

The selected topics represent by necessity my own particular academic background and interests rather than any attempt at comprehensive coverage. The range of topics, however, provides insight into the breadth of issues that are currently being tackled in contemporary neuroscience. Due to the quick evolution of developments in the field, a concern is that by the time you are reading this, the issue has ceased to be an "interesting problem"; such is the nature of a fast-growing science. In that case, my hope is that the historical analysis will still provide insight into how the topic came to be a problem of interest and lead the reader to an understanding of the contemporary solution.

I make no claim to being the first to notice this connection and my work here is indebted to the fine work of historians of science, psychology, and medicine. I am a historian by avocation, not formal training, and thus this book may be found wanting on that score. I am an experimental psychologist with an interest in the history of the field as opposed to a historian who may focus on psychology or medicine. While I often utilized the work of other scholars on a given topic, I also read original source documents when possible. I immersed myself in the literature of the topic during a given period so I could extract for myself the key points, the points of connection between then and now, and the differences in interpretation that come from varied socio-cultural and historical contexts. This might mean reading extensively on the development of scientific thought of a particular era, examining a discussion by physicians as reflected in medical journals over a period of years and decades, or tracing the development of a concept as illustrated in medical textbooks of different periods. Additionally, as a professor of several courses in a neuroscience program, I possess a working knowledge of the current neuroscience topics covered in this work, if not from an empirical research perspective, then a teaching perspective. Evaluating and summarizing contemporary research studies and analyses are a mainstay of my academic position in order to keep current in the field.

This is essentially a "then and now" book. I have selected topics that are both a sampling of current research in neuroscience and reflective of my own particular interests and expertise. Thus, there are more chapters that touch on psychoactive drugs than might otherwise be expected. The topics are by no means comprehensive and my great fear when I embarked on this project two years ago was that even the so-called contemporary research I discuss would be outdated by the time you read this. Keeping up-to-date in a field where there are breakthroughs in research on a weekly basis is a challenge.

The main focus of the historical analysis is the Victorian era, particularly in the United States and Great Britain, although some chapters warrant occasional discussion from earlier times or different regions. This emphasis is due to a number of factors. Given the broad array of topics, it was most efficacious to narrow the discussion to medical and scientific discourse among English-speaking regions; nervous system theories from

Western Europe have dominated science and medicine for centuries and the writings in the two regions of interest reflect those theories. This is not to suggest that there were no important ideas on the topics from other parts of the world but simply that they were not as well known or influential in shaping our modern views of neuroscience. In some instances, important early works by physicians in areas now known as the Middle East were readily adopted and disseminated by their European counterparts. On a practical note, I had greater access to primary source materials from these regions both because of language and physical proximity. I benefited from important sources of early medical and scientific thought originally written in Greek and Latin being available in English translation.

The book is organized topically, with each issue examined according to the contemporary theories, pertinent research, and an historical analysis. When examining a topic from an historical perspective, that sometimes means tracing the theoretical development chronologically with landmark events. At other times, a particular historical exemplar is highlighted for comparison. The lack of compartmentalization of academic disciplines or specializations in medicine means that writings on neuroscience come from a wide variety of sources from medicine, science, philosophy, and the early years of experimental psychology and physiology. My hope is that after reading a chapter, not only will your understanding of some of the "interesting problems" of neuroscience be enhanced but your curiosity about the brain and behavior will also be sparked further.

Introduction

"No one denies that the science of one generation is the pseudoscience of the next."—J.D. DAVIES, *Phrenology: Fad and Science*, 1955, x

THE NATIONAL "DECADE OF THE BRAIN" (1990–2000) and the "Decade of Discovery" (2000–2010) helped focus research and resources such that major advances in our understanding of neuroscience were achieved. Dr. Thomas Insel, director of the National Institute of Mental Health, pointed out in his 2006 address on the NIMH Strategic Plan that research on mental health issues has shifted from finding ways to decrease symptoms to understanding the brain mechanisms that underlie disorders with goals of prevention and cure.[1] President Obama in 2013 proposed a ten-year initiative for mapping the brain (BRAIN project); if adequately funded by Congress, the advances in this decade could potentially lead to both greater understanding of the neural networks that make up the brain and ultimately treatments and perhaps eventually cures of some neurological diseases.

As Jacyna and Casper indicate, the disciplines that include the neurosciences are ones that get at the very core of our human identity.[2] "Transdisciplinarity" rather than "interdisciplinary" is a term that has been used to describe the neurosciences.[3] Contributing disciplines include psychology, neurology, biochemistry, cellular and molecular biology, computer sciences, and philosophy—and this is by no means an exclusive list of scholarly fields! Because the unit of measurement ranges from the subcellular to the organismal level, the breadth of disciplines all bringing different perspectives to the workings of the mind and brain is noteworthy.

Yet this fascination, or "neuromania," is hardly new. Throughout the centuries scientific endeavors have yielded new insights that have both captured the public's imagination and led to genuine improvements in the treatment of mental and neurological disorders. However, these discoveries have always been the product of particular views of mind and brain that are formed and constrained by their sociocultural context. As these environments changed, so did the scientific theories. It may seem incredulous to the contemporary reader that there was at one time a question of whether the mind was situated in the heart but that view simply reflected the standard "scientific" dogma, as recorded by the ancient Greeks. Stanley Finger's *Origins of Neuroscience* contains an excellent and thorough accounting of the various theories of nervous system function through the centuries.[4] The purpose of this book is not to re-hash those theories as such but to juxtapose the assumptions and concepts with contemporary views. The changing landscape of medical historiography also warrants a re-analysis of assumptions. Contemporary neuroscience is especially guilty of taking an ahistorical stance as if historical context and culture had nothing to do with concepts now taken for granted.[5]

A number of recent works have sounded the call for skepticism instead of the unquestioning acceptance of neuroscience research.[6] Their foci have been concerned with methodological, philosophical, and technological assumptions of which the public—and even sometimes members of the academic community—are mostly unaware. Other researchers, such as Rose and Abi-Rached, make a case for neuroscience solving some of the most pressing problems of today—from addiction to psychopathology.[7] The potential for neuroscience providing significant advances in our understanding of individual and social phenomena is great. Both views have some validity.

The classic debate in neuroscience of whether the brain operates with distinct areas for specific functions or whether functions are distributed across many brain regions is the subject of Chapter 1. Functional Magnetic Resonance Imaging (fMRI) is one of the most significant technological developments of the last three decades. First developed as a more sophisticated and three-dimensional image of the body's anatomical structures, the ability to reveal metabolic processes within the brain *as they happen* is what has led to the research explosion in neuroscience and its various offshoots, e.g., neuroeconomics, neuroaesthetics, and neurocognition. However, it turns out that the technology is not so straightforward. Amazingly, much of the critiques and lauds for fMRIs in neuroscience parallel the commentary that greeted the beginnings of phrenology. Now firmly relegated to the bin as a pseudoscience, phrenology was once the cutting-edge technology of brain science. The enlightenment proponents and critics of the early 1800s centered on the same localization versus distribution debate that exists today. Phrenology and its modern-day rendition, fMRIs, are discussed in order to examine the metaphor and ascertain its validity. Perhaps better than any other topic, this metaphor reveals both the uncertainties and the promise of neuroscience.

In Chapter 2, I examine the idea of "the double brain," the notion that we have two functionally separate brains manifested in our two cerebral hemispheres. While we generally accept today that we have two hemispheres each having some specialized functions, just how those hemispheres worked together or in parallel was the subject of some debate in the 19th century. Did we, in fact, have two independent brains in our skulls? Victorian physicians believed that many poorly understood mental phenomena, such as hypnotism, somnambulism, and what we would now term dissociative disorders (e.g., fugue states and multiple personality disorder), could now be explained by having such a double consciousness. The emerging field of neurology in the second half of the 19th century also examined the concept to explore the biological bases of aphasias and agnosias. There was then a quiescence of research until the 1960s when epileptic patients underwent a surgery to have the fibers connecting their two hemispheres severed in order to control their seizures. This "split-brain" procedure reignited discussion of some of the very same issues. The key difference now was having an experimental methodology to examine and confirm the hemispheric specializations rather than relying on neurological case studies.

Chapter 3 is an investigation on that ubiquitous behavior, sleep—a behavior that was poorly understood until the modern era. In fact, it was considered almost a non-behavior or absence of behavior by psychologists and medical scientists until the accidental discovery in 1953 of rapid eye movement sleep.[8] Sleep was simply the absence of consciousness, while now considered an altered state of consciousness. Even today we can say that we can better describe what goes on physiologically during sleep, what sleep "looks like" in terms of electrical activity that make up brain waves, what brain structures are involved yet much about sleep remains a mystery. What triggers it, for example? What are the underlying causes of the plethora of sleep disorders or dyssomnias, many of which were unheard of 30 years ago? A perusal of the history of sleep shows that the questions of today mirror the questions of yesterday. From early theories of blood congestion to nerve relaxation to the supposed presence of hypnotoxins, we see the continuation of many of the same themes throughout as we do today in contemporary research. The relationship of sleep and consciousness was a focus then as now.

Music has long enjoyed since recorded history a reputation of being beneficial to one's health and in recent decades has been the subject of an ever-growing area of neuroscientific research. There is much interest, both on the part of the public and academics, on how the brain perceives music and, conversely, how music affects the brain. In Chapter 4, this relationship is examined by exploring contemporary neuroscience research. However, in looking at its historical counterparts, the focus is on what I consider to be the more interesting question: When is music bad for one's health? One can readily find evidence throughout medical history on the positive attributes of music and its role in healing; but there were also periods, particularly during the Enlightenment era or "long 18th century," when certain types or aspects of music were thought to be deleterious to one's health. The special relationship of music and the nervous system is also explored in this chapter, focusing on the sensibility and resonance of nerves. Indeed, excessive exposure to particular types of music or even instruments, such as the glass harmonica, could risk descent into madness! Nervous music was a potentially serious problem, especially for females with greater sensibility, and health treatises cautioned the listener on the liabilities of the wrong kind of music. Lest one think that was an isolated historical anachronism, only consider the initial reaction to Wagner's music in the 19th century and jazz and punk rock in the 20th.

Chapter 5 explores the quest for altered states of consciousness by the use of psychoactive drugs. These are the drugs that cross the blood-brain barrier into the body's nervous system. Throughout history, people of virtually all recorded cultures have used substances to change their emotions, thoughts, and behaviors; some of these substances occur naturally in the environment while others are more recent lab-synthesized creations. Some of the current illicit drugs are new, many are as old as time, yet the use and abuse of psychoactive drugs has been a constant. With the development of cheap mass media via the printed press came drug advertisements, celebrity endorsements, and scandalous exposés. These all helped variously to glamorize and demonize psychoactive drugs. This chapter will focus particularly on the drug exploits and explorations of some well-known, late Victorian experimenters.

The accidental discovery of LSD by Albert Hoffmann in the 1940s is a well-known story in drug lore. Lesser known is the experimental research on hallucinogens that occurred before Hoffmann's time. In Chapter 6, I expand on the research discussed in the previous chapter by examining in more detail the hallucinogenic research that was ongoing during the 19th century. Another piece of hallucinogenic experimentation that tends to be overlooked, however, is the research program using LSD and mescaline for therapeutic purposes. I follow this story in order to show how a once promising treatment for conditions like alcoholism was curtailed and ultimately villainized. That research program came to a grinding halt as a consequence of new federal drug regulations enacted in the 1970s that made scholarly research on psychedelics nigh impossible. However, there has been a bit of a come-back of these drugs of late on the research scene. Recent relaxation of the regulations has led to new drug trials of hallucinogens for therapeutic use under tight controls.

Chapter 7 takes the psychopharmacological discussion a bit further by concentrating on opium particularly during the mid to late 1800s and contrasting its usage with the current epidemic of prescription pain pill addiction. The history of the use and abuse of opium is especially enlightening for understanding some of the current debates on addiction and addiction therapies. Once considered the most essential mainstay of the physician's pharmaceutical toolkit, opium and its derivative, morphine, were instrumental in social upheaval and even the cause of war. None of these opiate drugs were originally illicit. Indeed, these drugs were used medicinally for centuries as analgesics, stimulants, and cough suppressants, for example, until societal pressures were exerted to criminalize their use. Rarely were they supplanted by safer, more effective medications; more often the substitutes came with their own set of problems. The dialogue concerning opiate use during the late 1800s and early 1900s by the medical community contains many of the same arguments still used today about fears of addiction with palliative care. Theories about inebriety, insanity and nervous system function all had a role to play and culminated in a basic philosophical division still seen today: Is addiction a disease caused by a malfunctioning brain or is it the manifestation of weak will?

One of the most interesting developments of late in the search for both the cause and a cure for schizophrenia has emerged from research examining the relationship between the brain and the gastro-intestinal tract. The gut's microbiome has been shown to interact with brain function in reciprocal and sometimes dramatic ways. In Chapter 8, past and present research on the link between brain, diet, and gut is explored. From ancient times, physicians have utilized a variety of remedies to purge the gut in the interest of promoting good health or cures. The close relationship of diet and health was assumed by all and it was the rare medical text that did not include extensive discussion of foods or dietary prescriptions to promote health or healing and those to consume at one's peril. Less accepted was the relationship with the nervous system and it was only relatively recent that the physiological connections between digestive and nervous systems were established. However, one of the most popular patent medicine remedies of the 19th century was a dietary pill that was purported to cure any illness but most especially nervous disorders. This medicine was Morison's vegetable pills, the key ingredient for James Morison's Hygeian system of health. Morison constructed a belief system for his customers and a powerful purgative with his pills. Purging the body of toxic chemicals was supposedly achieved by this concentrate of pure vegetable matter and the result was improved health—except, of course, when the result was actually worsened health or even death. The remedy bore all the hallmarks of quack medicine and those characteristics are examined both in the case of Morison's pills and modern dietary detox regimens.

Searching for medical or pharmaceutical aphrodisiacs has long been an industry quest and an area ripe for exploitation. Countless remedies from ginger to Spanish Fly have been recommended from the ancient Greeks and Romans to the modern era. Significant developments in the treatment of erectile dysfunction did not occur until first with "rejuvenation" therapies in the 1930s and then in the 1950s and '60s with vacuum pump technology. The discovery that led to the development of Viagra (sildenafil) has been well documented elsewhere almost to the point of being transformed into hagiography by Pfizer pharmaceutical company.[9] Nonetheless, the discovery of sildenafil's action on erectile tissue was a serendipitous development from the Nobel Prize–winning research on nitric oxide and the cardiovascular system; initially, there had been no thought of application to erectile dysfunction. Chapter 9 examines the medicinal fads in the treatment of impotence, the relationship with male hysteria, and assumptions about neuro-sexology, the relationship of the nervous system and sexual performance, then and now. Viagra was a solution in search of a problem and highlights the interplay between Big Pharma, psychology and medicine; the transformation from "impotence" to "erectile dysfunction" is a striking example of a generational shift in medical discourse and transfer of ownership of a disorder. The FDA approval of "the pink pill," Flibanserin —the "female Viagra, " has put this issue back in the spotlight.

The perennial fascination with psychological types is the focus of Chapter 10. Typologies have been popular since ancient times with Galen's famous theory based on humors in the body: water, phlegm, black bile, and yellow bile. The idea that personality and, pertinent to our discussion, nervous systems, can be categorized into discernable types is resurrected periodically both by the popular lay writers and in academic circles. It is a compelling idea, attractive in its simplicity and promise for easy explanations. However, the ease of understanding individual differences often comes with a price of logical tautologies. These are the same fallacies that make trait theories vulnerable. An individual is shy because of a trait of shyness; likewise, an individual who has a trait of shyness will be shy. In fact, there is no explanation at all for shyness but just a circular argument. Similar to typologies, creating categories based on different patterns of nervous system action or bodily manifestations simply re-labels the descriptive phenomenon. This is much like behaviorist psychologist B. F. Skinner's notion of "unfinished causal sequences"—using terms that themselves need explanation to explain phenomena.[9] The latest manifestation of biological typologies takes the form of trait theories, usually with a genetic underpinning. Even contemporary personality theories take their lead from genetic endowment. The connection with neuroscience arises in discussions of addictive personalities, for example, once thought to be a function of constitutional weakness and more currently a result of bad or defective genes. Similarly, the interest in detecting a neurological flaw that can reliably reveal the psychopathic or antisocial personality represents something like the "Holy Grail" of forensic psychology. Again, an historical analysis of the problem can help reveal the pitfalls of such simplistic thinking and the avenues which might provide insight.

Chapter 11 focuses on a popular 19th-century treatment for alcoholism that laid the groundwork for contemporary programs today. The treatment had elements of quackery, particularly in regard to the pharmacotherapy that was the signature feature of the program. However, the complete program incorporated supplemental treatments using social support, structured rehabilitation, and follow-up procedures to promote maintenance of abstinence. The drug used for detoxification, available both in-patient and by mail order, was Keeley's Bichloride of Gold. The treatment program was established in the Keeley Institutes, a franchised operation that was a complete detox and rehabilitation program for alcoholism, morphinism, or tobacco addiction.

Detox programs of the 19th century operated with the assumption that drug addictions reflected brain abnormalities particularly in those individuals prone to degeneracy. The model of addiction as a moral weakness that was prevalent in the preceding centuries had given way to a more comprehensive "addiction as disease" perspective based on the newer theories of nervous system functioning. This model would then wane as psychiatrists followed the path of Freudian talk therapy and neurologists failed to find organic abnormalities that presaged addiction. In the modern age of neuroscience, drug cures for drug problems are being attempted once again. The disease model of addiction is back and promises for a chemical cure along with it.

In Chapter 12, I explore the topic further by examining one of the most popular medical treatments for neurasthenia developed in the 1870s by Dr. Silas Weir Mitchell. Called variously the "Rest Cure" or the "Weir Mitchell Cure," it achieved notoriety from the 1892 quasi-autobiographical short story *The Yellow Wall-Paper* by Charlotte Perkins Gilman.[10] While the Rest Cure can be viewed in the context of contemporary drug and alcohol detoxification programs, it was not intended as such. Its creator, Weir Mitchell, was one of the preeminent American neurologists of his day, with an impressive list of contributions to the fields of neurology, toxicology, and experimental physiology. However, the Rest Cure is his most enduring legacy, not because of its medical merits but rather by being vilified by literary feminist scholars in the 1970s based on Gilman's fictional account. In this chapter I review Weir Mitchell's actual work in the context of accepted medical knowledge and practice of the time in order to evaluate the treatment. Methods for treatment of neurasthenia varied greatly in late Victorian times, especially dependent on whether the attending physician was operating from the neurological or psychiatric perspectives—all during a time when the division between those professional subspecialties was blurred. Mitchell's treatment, as outlined by himself and corroborated by other leading physicians at the time, is described and is then compared directly with the account in Gilman's short story. This is important because even today *The Yellow Wall-Paper* is included in the contemporary canon read by thousands of literature students as an example of Victorian male oppression by defining feminine initiative as insanity. The details of the treatment, while secondary to the literary function of the story, matter because it is the treatment itself that is the tool of the masculine oppression. The points of difference do not detract from the literary merit of the work but instead help enlighten one about perceptions of medicine, health, and wellness at the time. Reading the details of Mitchell's Cure reveals a striking similarity to contemporary "detox" programs; given the heavily medicated patients of the time, it is conceivable that many of the confusing symptoms of neurasthenia were, in fact, signs of overmedication with psychoactive drugs.

These topics represent a sampling of issues that have fascinated the public for as long as the nervous system has been identified. While our knowledge of neuroscience has increased manifold over the last century, the nature of the problems and often their solutions have changed very little. It may be inherently interesting to learn about the history behind some of these topics but our understanding of these phenomena is also enhanced because often assumptions are rooted in the historical development of the concept. In all instances, they have led the way to advances in our knowledge—sometimes for the betterment of society and therapeutic breakthroughs and sometimes not; sometimes with a straightforward, linear path of discovery but just as often research dead-ends and mistakes. More often, supposedly new problems are just the reframing of old problems or being examined with the latest technological advances in measurement and assessment. Either way, our understanding of present day neuroscientific questions will benefit from the knowledge that comes with historical perspective and reasoning.

The Brain

1. The New Phrenology

One of the classic debates in the history of physiological psychology is the localization of brain function. Where are particular abilities, talents, and personality propensities located in the brain? Localization versus distribution of function continues to be a major issue in contemporary neuroscience as well, with much basic research still being done to isolate pertinent brain areas involved with particular functions. Modern imaging techniques have led to exciting developments in establishing the role of various brain structures. Yet this very technology still raises some of the fundamental questions posed during the last 200 years or so about the workings of the brain without resolution.

Much has been written on the spirited 19th-century debates between localization and distribution of psychological functions in the brain but nowhere is it in sharper relief than the early discussions centered on the validity of phrenology.[1] While phrenology has long since been relegated to the barrel of discarded pseudo-scientific theories of the past, more recently its name has been invoked in regard to a contemporary trend in psychological and neuroscience research: functional magnetic resonance imaging or fMRI. I am by no means the first to make a connection between phrenology and the use of fMRIs, nor to express caution against the frequently over-generalized claims of fMRI psychological research.[2] Because both techniques rely on an assumption of localization, the connection between the two is an obvious one—at least to those cognizant of the history of the debate. But while some have questioned the claims of fMRI research outcomes, eloquently spelled out the constraints in interpreting fMRI studies, and even spoofed the connection, none have gone so far as to explore the analogy fully to see where the connections are valid and where they are misapplied.[3] The purpose of this chapter is to take on just such an investigation. In doing so, I will give brief overviews separately of phrenology and fMRI, examine their assumptions, contributions and weaknesses, and then address the points of similarity, as well as where the analogy fails to get traction.

Franz Joseph Gall is credited with establishing phrenology in the 1790s, although organology was his favored term.[4] He was an accomplished and well-respected anatomist, known for developing a new dissection technique that isolated fiber tracts vertically from their nuclei. With that technique, he demonstrated that the continuity of nerves as tracts in the brain and followed those tracts to their respective points of origin. He then proposed a theory of brain function that had as its fundamental assumption the notion that mental phenomena have natural causes, i.e., mind equals brain. Not only was the brain the seat of intellect, sensation, and volition—a point that was not yet fully agreed upon by the scientific-medical establishment—according to Gall, one could infer aspects of character or personality in the physical representations of the brain.

Gall famously wrote that he derived his ideas as a child from his observations of his peers. One lad who had bulging eyes was notable for his exceptional memory. The physical characteristic was thus correlated with a psychological phenomenon.[5] In order to avoid the heretical charge of materialism that followed from his public lectures (plus the fact that Emperor Francis I banned his lectures), he left his medical post in Vienna to travel the European continent with his student, Johann Spurzheim.[6] As Harrington states, he was among the first anatomists "to map out the human soul boldly upon the convolutions of the cerebral hemispheres."[7] Materialism was considered heretical because it implied that the mind was simply matter, rather than an immaterial soul. Gall did not himself consider this theory as being counter to the Church because one of his proposed faculties or "organs" was religious sentiment. Having an anatomical organ to know God was proof for Gall of God's existence.[8] Gall's emphasis was on the facts of anatomy as he perceived them. He eventually settled in Paris where he spent the remainder of his life and there promoted his organology through lectures and publications despite negative critiques from the prestigious French Institute.

Contrary to the prevailing view in the 18th century that the brain constituted a "common sensorium," Gall proposed that the brain had 27 independent and measurable organs that correlated with mental faculties.[9] These faculties ranged from Cunning, Pride, Mimicry, and Ownership to Amativeness, Poetic Talent, and Religious Sentiment. He based his assertion on evidence obtained by a variety of methods. He collected an array of human skulls and head casts for study from a variety of sources, as well as animal skulls for cross-species comparison. Many of these skulls were from individuals that had exhibited some remarkable characteristic during their lifetimes such that Gall was able to compare extreme examples of a particular faculty. Likewise, in his travels, he sought out exemplars of unusual abilities—both deficits and accomplishments—in living persons or similarly, individuals with unusual physiognomy so that he could interview them about their talents and personality characteristics.[10] Because he believed that the cranium faithfully represented the brain matter beneath it, he did not typically examine the brains themselves when available.

It was his student and partner, Johann Spurzheim, who promoted and extended the theory of organology far beyond the physiological claims that Gall had originally proposed. Actually, it is due to Spurzheim that the term "phrenology" came into use and the theory was popularized.[11] Gall eventually split with Spurzheim over differences in theory, such as the emphasis on psychology rather than physiology, and their former partnership was permanently ruptured. According to Finger, Gall had been interested in the anatomical implications for organology while his colleague was more interested in the practical application of phrenology.[12] Spurzheim was also more willing to delve into metaphysical arguments and "looked forward to the perfection of the race by the aid of phrenology."[13]

Spurzheim expanded the number of faculties from 27 to 35 and reconceptualized them in such a way as to be consistent with the Scottish Faculty psychology of Thomas Reid and Dugald Stewart. This was ironic in that the "Scottish school of commonsense," led by Reid and Stewart, was vociferously against phrenology. Their faculties were "reflective categories or processes of the mind at work rather than places in the brain."[14] The cerebrum was still widely held to be a unitary matter and indivisible. Lange described the phrenology perspective as "a parliament of little men together, of whom, as also happens in real parliaments, each possesses only one single idea which he is ceaselessly trying to assert.... Instead of one soul, phrenology gives us nearly forty."[15] Spurzheim traveled extensively in England, Scotland and Ireland where his ideas were well received. Negative faculties were replaced with more positive ones. He also saw the potential for application of the theory to progressive education and self-improvement, an idea that particularly appealed to the American audiences.[16] While Gall was the founder, Spurzheim was the evangelist for the movement.[17] Indeed, many of the roots of various social progressive movements in the United States focusing on self-improvement, schooling, reform, penology, and treatment of disabilities, can be traced to phrenological principles. Indeed, Spurzheim's expansion of the theory behind phrenology was to create an organization that mimicked the social class structure and gave direction for self-improvement through the ranks.[18] Some researchers suggest that this was a key sticking point for critics of phrenology, that the theory went against the status quo.[19]

A Scottish lawyer, George Combe, became enamored of phrenology and took the applications even further in his 1827 best-selling work, *The Constitution of Man*.[20] He co-founded the Edinburgh Phrenological Society with his brother, as well as its publication, *Phrenological Journal and Miscellany.* Societies in other cities in Britain and the United States flourished and Combe became the leading spokesperson for phrenology. In the 1830s *The Constitution of Man* was the third best-selling book in the United States, outsold only by the *Bible* and *Pilgrim's Progress.* And yet by the late 1830s, phrenology was all but finished as a scientific endeavor.

The basic principles of phrenology were as follows:

1. There exist innate faculties.
2. The Brain is the organ of the mind. Until Gall's definitive anatomical and physiological work, this point was still under debate. Gall's research settled the question and this is probably the most important contribution of phrenology to modern day neuroscience.
3. The Brain is an aggregate of organs rather than a homogenous structure. This was central to the contemporary debates in the early 1800s.
4. The cerebral organs are topographically localized. In other words, there is a specific location for each organ of faculty.
5. The skull manifests organ development due to ossification during infancy. This was the crux of the craniology theory. Since ossification of the skull was incomplete at birth, the skull would then conform to the underlying brain organ.
6. Exercise of the faculty results in brain organ development.[21]

The basic paradigm for the phrenologists can be outlined thus. A striking talent is observed in an individual and that presumes the existence of an innate faculty. The faculty represents the function of a cortical organ. The workings of the cortical organ in turn is manifested by some degree of cranial prominence relative to individuals without such a pronounced talent.

The reasons, finally, why phrenology was ultimately rejected by the scientific community were manifold. One reason is known as the "Doctrine of the Skull." This referred to the notion that the skull reflected the underlying cortical prominence. Scientific evidence attacked this principle in two ways. There apparently was no anatomical relationship between the shape of the skull and the shape of the cortex, certainly not in any way that corresponded with the 27–35 faculties.[22] Likewise, there were no actual anatomical delineations of organs corresponding to the faculties on the cortex.[23] Many scientists who were favorable to the notion of brain influencing character reluctantly had to acknowledge the lack of data.

Secondly, Marie-Jean-Pierre Flourens' work in Paris on mass cortical action seemed to disprove any theory of brain function localization.[24] A stalwart of the French medical establishment, he was one of Gall and Spurzheim's chief critics who sought to discredit phrenological theory both on the grounds of physiology and theology.[25] While Flourens believed that certain brain structures, such as the medulla, did have specific functions, he did not extend this belief to the cortex, which he thought was a unitary substance. His experimental work on the cerebellum decidedly showed that it was a structure involved with coordinated motor behavior rather than amativeness, as the phrenologists claimed.[26] Using a variety of animal subjects with a careful experimental method, Flourens was able to show that ablations (removal of tissue) in a variety of cortical locations resulted in the same functional deficits. In other words, there was no differentiation of function as a result of ablation or stimulation of different areas of the cortex. The cortex exhibited equipotentiality.[27] This idea turned out to be incorrect as Flourens had used animals with less developed cortices and very elemental behaviors. As Stanley Finger notes, Flourens' methods were sound but his conclusions were flawed; the opposite was true to some extent among the phrenologists.[28] Localization of function at the cortical level would be demonstrated a few years later by other French physiologists and physicians (e.g., Paul Broca) in clinical case studies. Even William James in his *Principles of Psychology* admitted that while phrenology had an attractive premise and that localization of some cognitive functions had since been demonstrated, the methods to examine faculties used by phrenologists were "vague and erroneous from a psychological point of view."[29]

Methodologies also presented problems for phrenologists. They relied heavily on case studies, particularly cases of the extraordinary rather than the average, anecdotal evidence and correlational data. Devotees actively sought out confirmation of their theory and so cherry-picked the data. The gravest issue was the lack of falsification for the theory. Contradictory results could always be explained such that no data were sufficient to ever disprove the theory—a serious flaw to any theory purporting to be scientific in nature. There was an explanation or excuse for every inconsistency. For example, an under-developed organ might still yield a strong showing of a given faculty due to the compensation of other areas.[30] Besides the complete arbitrariness of the selection of 27 faculties, explanations were also hampered by tautological or circular reasoning, a problem of many faculty or trait theories. As Young states, "they substitute classification for explanation."[31] Fancher notes that "as long as phrenology lacked an adequate system for describing psychological characteristics, it could never hope to account adequately for differences in human personality."[32]

A third reason for rejection, however, was the migration of phrenology from the medical/scientific community to the popular or vulgar masses. The "vulgarization" of phrenology with its applications for personal self-improvement, implications for liberal social policy, and worse yet, its appeal to entrepreneurs, sealed its fate as a former science now relegated to the cupboard of pseudoscience.[33] "Practical Phrenologists" who began commercial enterprises emerged, especially in the United States, and were differentiated from the physicians who had incorporated the concepts of phrenology into their medical practices.[34] Van Wyhe (2004a) contends that phrenology had popular appeal precisely because it gave the individual more authority of what was "truth" concerning one's character.[35] Itinerant phrenologists in the mid–1800s were "bastardizing ... established science."[36] The Fowler family (Orson, Lorenzo, and Charlotte; Samuel Wells) exemplified the growth of the entrepreneurial phrenologists and was responsible for its resurgence in mainstream popular culture in the late 1800s.[37] The American Phrenology Institute alone by 1894 had 600 graduates who had each paid $50 for a course in becoming a practicing phrenologist. Amazingly, Charlotte Fowler was admitted as a fellow to the American Association for the Advancement of Science (AAAS), one of the most prestigious scientific organizations in the United States.[38]

Nonetheless, it is important to realize that between the 1790s and 1840s, phrenology was very much mainstream medicine. If one examines individual issues of *The Lancet*, the premier medical journal in the United Kingdom then as now, one will find numerous articles, letters to the editor, editorials, and reprinted lectures all firmly in support of the phrenology doctrine with very few opposing views.[39] Lectures by Spurzheim and subsequent phrenologists were often reprinted in full. Phrenological principles were taught in the leading medical schools during those decades as well. In Boston alone, more than half the physicians were favorably disposed towards phrenology in the 1830s.[40]

Yet by 1830, the Boston Phrenological Society ceased its operations.[41]

Internal dissent among phrenologists themselves regarding the number and nature of faculties and how one should apply the knowledge hastened the demise of the movement overall. One division concerned the secular application of phrenology. Combe's *Constitution of Man* was clearly in the secular camp of demonstrating man's agency versus older "evangelical" approaches.[42] An 1842 report in *The Lancet* notes a split in the British Phrenological Society due to the "mesmeric clique," alluding to some tendencies to delve into areas of questionable scientific reputation.[43] John Elliotson, a British physician best known for introducing mesmerism (i.e., hypnotism), as well as the use of the stethoscope to British colleagues, had founded the Phrenological Society of London, which he later resigned in order to form the London Phrenological Society in 1838.[44] Forays into mesmerism and spiritualism tarnished the former glow of phrenology, leading not only to an early split among phrenologists themselves but increasing skepticism by the medical establishment.[45] However, Elliotson continued to have his lectures and reports printed by *The Lancet*.[46]

The proportion of favorable to unfavorable reports clearly shifted during the 1830s and by 1851, there is the first major article referring to phrenology as a pseudoscience and quackery.[47] The early criticisms of the theory were based more on philosophical grounds, charging materialism, rather than on physiological foundations. Only towards the end of this period did the physiological and anatomical evidence in refute of phrenology begin to make an impact. In the meantime, the popularization of the field was associated with phrenology degenerating into quackery: "bumpology." Even in 1824, there were calls from the medical journals to distinguish "Bumpists" from phrenologists.[48] Yet in 1836, *The Lancet* printed 20 lectures on phrenology by Dr. M. Broussais, who declared the field to be "the physiology of the brain."[49] The year 1838 is the first volume for *The Lancet* in which there is no mention at all of phrenology. Subsequently, articles relating to phrenology are fewer and critical editorials, especially in regard to methodological flaws, more the norm.

An excerpt from the *Boston Medical & Surgical Journal* on December 27, 1843, is instructive: "Phrenology, after undergoing a variety of degradations by being mixed up and compounded with the rarishow exhibitions of animal magnetism, will by-and-by be resuscitated and shine with its former splendor; and will one day have a place with the exact sciences, where it legitimately belongs, but from which it has been kept away by the enemies of its discoveries."[50] Phrenology had taken both a dogmatic and ultimately a methodologically flawed approach that would eventually doom itself as a serious contender for legitimate science. The field had a resurgence of popularity among the masses, if not the scientific community, in the late Victorian era and phrenology still occasionally rears its head in the modern era.[51]

Now on to the 20th century and functional magnetic resonance imagery (fMRI). First, I want to distinguish between MRIs, which provide high resolution images of the brain and used as a now standard diagnostic tool for examining anatomical structures in the brain, as well as other organs. In contrast, fMRIs provide images depicting which structures are activated during different tasks or activities, in other words, giving a glimpse into the underlying physiology of neural function. While also used in the aid of medical diagnoses, fMRIs are increasingly used as a major research tool in cognitive psychology and neuroscience. Interestingly, the technology relies on a paradigm very similar to that of phrenology: using visual representation to infer brain function in investigations of the mind-brain relationship. However, whereas cranial prominence was the main feature of phrenology, increased metabolic activity as depicted by a certain color on constructed images is the focus of the fMRI. There is an assumption that increased levels of oxygenation, reflected by increased blood flow to a particular area, are associated with increased neuron action potentials (neural information) and thus the observer can infer that the structural area of interest is likewise activated.

There are a number of distinct assessment techniques used in neural imaging but one of the most common utilizes the BOLD signal: blood oxygenation level-dependent.[52] When the individual's head is placed in a strong magnetic field, oxygen molecules align themselves with a particular pole. This slight movement is detected by the scanner and the difference between oxygen-rich and oxygen-poor areas is determined. Neurons in action are high users of energy and require increased glucose, which in turn means increased blood flow that carries oxygen with it. Like other measures of metabolic activity, the BOLD signal is an indirect measure of neural activity and one based on relative differences between areas. However, it has advantages of being non-invasive, not requiring radioisotopes, and having superior spatial resolution over older methods, such as positron-emission topography (PET) scanning.[53] Alternative techniques used today include voltage-sensitive dye imaging (VSDI), arterial spin labeling (ASL) and diffusion tensor imaging (DTI). The latter is an especially important new technology because it allows for the tracing of neural axons or tracts. This means that neural pathways can now be followed, which advances our understanding to a higher level of circuitry and modules beyond simply what structures are activated or depressed during particular tasks.

fMRIs, however, have unique underlying issues and assumptions and as a widespread research tool that is approaching the 20-year mark, those assumptions can often be overlooked and certainly invisible to the casual consumer of scientific news. Wager et al. describe the limitations of the fMRI in three categories: temporal, spatial, and artifacts of acquisition.[54] This chapter will only touch on the most critical assumptions and characteristics of fMRIs that might affect interpretation of mind-brain functions and are relevant to the charge of phrenology redux. Newer technologies for assessing brain function, such as arterial spin labeling (ASL) and diffused tensor imaging (DTI) are not included in the discussion.

One of the most important characteristics of fMRIs for research concerning attributions of brain function is the well-known temporal lag issue.[55] While the imaging produces high quality spatial resolution, it comes at the expense of temporal resolution. The fMRI does reflect potentially second-by-second changes yet the hemodynamic process in the brain takes about six seconds compared with two *milliseconds* of a neural impulse. That is a considerable time in neural terms to measure brain activity, especially when compared to the real-time recordings of neurons by electro-encephalogram (EEG). The signal may also vary because of "slow drift," which impacts the ability to study functions that may take longer to process (e.g., emotional responses). Is the result because of the artifact or the process, in other words? The inferior temporal resolution makes it difficult to couple hemodynamic changes to discrete neural events but averaging across multiple trials helps. Many studies now combine EEG recording with fMRI in order to improve upon the temporal resolution.[56]

Similarly, there is an issue of too many false positives altogether. The technique is particularly vulnerable to certain types of false positives because of the multitude of statistical tests, since many trials have to be performed in a given experiment. A well-known axiom in statistics is that the more tests that are run, the more likely one is to find significant results. An infamous study demonstrated this when Bennett and colleagues presented a poster on a cognitive ability of a salmon.[57] There was the typical fMRI scan purporting to show the increased activity in a particular brain area as proof of the localization of the salmon "thinking" about a topic. The only problem was that the salmon was, in fact, dead. Simply running the multitude of statistical tests on the many "slices" of the salmon's brain yielded some statistically significant results. In a more serious article, Bennett argues persuasively for utilizing the appropriate statistical controls for multiple comparisons.[58] Because fMRI studies typically require multiple trials in order to weed out noise and artifacts like head movements and respiration, the statistics require sophisticated methods of control. These studies also lack statistical power due to very small sample sizes, thus limiting inferences that can be made from results.[59]

The spatial resolution of fMRIs can be remarkable, albeit not suitable for very small areas of the brain. However, it is also important to realize that the scanning images reported in scientific journal articles (and often reprinted in the popular media) represent average activity of the brains in a particular sample. There are considerable individual differences in brains with respect to size and locations of structures, sometimes as much as a few millimeters difference. That may not seem like much of a difference but when one considers that there are millions of neurons contained within those few millimeters, the standard error in averages may be marked.[60] Technical procedures called warping have been developed to better account for these differences but in the process, may cause blurring of the spatial resolution or false positives.

Technical decisions of both image resolution and slice thickness can impact results powerfully. One way can result in too many false positives—indicating changes in activity when in fact there is none; another can result in missing activity altogether. The particular unit of measurement is typically a voxel, basically a volumetric pixel, and the scanning will focus on that three-dimensional space with the million or so neurons contained within it. The specifications are determined by the researchers and may vary from study to study, thus making it difficult to compare experimental results. Researchers can test hypotheses using a Region of Interest (ROI) that at least focuses on particular areas thought to be involved with a brain function. Still the number and size of the ROI can affect the outcomes. The ROI must contain a sufficient number of neurons to pick up the change in activity. "Many regions may show changes in neural activity that is missed because they do not change the net metabolic demand of the region."[61] Even single voxels may contain subsets of neural circuits where the activity would be missed amongst all the neurons within that voxel. This type of imagery is not especially good at determining micro-circuitry among small brain nuclei. Also, when a limited region is involved in a variety of brain functions, as indeed most are, inferences about localization of function become problematic.

There are also a number of experimenter choices which affects results. These range from the choice of colors that are selected to represent more or less activity, what constitutes a threshold for determining whether a voxel is "active," to decisions regarding what activity is "noise" versus "signal."[62] The early research was plagued with artifacts masking positive results, to include jaw clenching in participants and the "slow drift" of signals. Better research protocols now control for most of these artifacts.

Early research was also published using poor experimental designs and even today there is a tendency, especially among writers in the popular media, to imply causation when the study is not actually experimental (i.e., variables are not being manipulated in a controlled fashion). While most current BOLD studies utilize a design called the subtraction method (condition A + B, followed by condition A, in order to see what effect B had), the design still flaws because it assumes that the change in condition is not accompanied by any other changes in the brain outside the ROI.[63] Sample selection bias is a problem in that the study participants are typically not randomly selected or even randomly assigned to conditions. Some individuals cannot tolerate the scanning process with its claustrophobic environment, loud banging noises, and issues with magnetic or radio frequency interference. Some people (especially children) are incapable of holding still for the 45 minutes or so for a typical scanning session. The experimental tasks done while scanning have often lacked real world transferability. While new apparatuses are continually being developed, such as virtual reality glasses that can be worn within the scanner, many studies have to have tasks that lend themselves to a fixed head confined in a relatively small enclosure.

Thus, it should be apparent by now that fMRI studies are not infallible in their conclusions. The search for localization of function in the brain has progressed considerably in the last two decades but there remain kinks in the process to be worked out satisfactorily. The most progress arguably has been made in the area of structure-function mapping, which has been largely exploratory and descriptive rather than inferential or hypothesis-testing. The latter, with its implications for practical applications has yielded intriguing but often conflicting results. Satel and Lilienfeld present an excellent overview of how imaging results have been overstated with respect to applications.[64]

So how does the use of fMRIs compare to phrenology then? The basic paradigms that serve each methodology are quite similar. Both approaches involve internal cognitive functions whose location relies on the underlying brain structure which will manifest itself in some visual manner, whether by cranial prominence or increased metabolic activity. For phrenology, the paradigm went something like the following. First, a striking behavior would be observed that was indicative of a talent or special ability. This talent resulted from an innate faculty and was reflected by the functioning of a cortical organ in the brain dedicated to the function of that faculty. The enhanced functioning led to greater cranial prominence in that area as indicated by physical size. In a parallel scheme, the paradigm for fMRIs can similarly be construed. Again, a striking behavior that is indicative of a particular skill or ability is investigated. This skill or ability is an output of some underlying cognitive function. That cognitive function, in turn, is localized to one or more specific brain structures and is manifested by greater neural activity in those areas.

For phrenology, the paradigm can be sketched schematically as follows:

Talent	→	Striking Behavior
Innate	→	Faculty
Function	→	Cortical Organ
Size	→	Cranial Prominence

In a parallel scheme, the paradigm for fMRIs follows a similar format:

Talent	→	Striking Behavior
Innate	→	Cognitive Function
Function	→	Brain Structure
Size	→	Increased Activity

The similarities are not entirely parallel as phrenology relies on the physical properties of the brain structure, while fMRIs rely on the physiological properties. This difference is expressed in the last blocks of the two schematics: greater cranial prominence versus increase activity. However, the former is contingent on the latter. Both are based on what is/was believed to be the currently accepted anatomical structures and nervous system principles of the day and propose to measure the generalized mind. There is a shared assumption between the two technologies that anatomical structures are only involved in singular functions. While in contemporary neuroscience, we know that is not actually the case, rarely does one see that explicit caution in interpreting results of a study. The resolution of that discrepancy may lay in the activity of subnuclei that have neural activity too small to be yet detected or among the multitude of pathways among structures and various neurotransmitter receptor subtypes.

Both technologies relied, at least initially on similar methodologies. Case studies and correlational studies were and are favored approaches and neither research methodology demonstrates causal effects. These approaches are valued for rich descriptive data, exploratory research, and generating hypotheses; however, they have severe limitations in testing hypotheses. Specifically, the types of experimental controls necessary to demonstrate causation are simply not feasible with those technologies. It should be noted, however, that better research designs in recent fMRI studies are giving rise to data that are more reliable and from which better inferences can be drawn.

Experimenter bias is a potential problem in both technologies. Rarely are studies performed with the experimenter blind to the participant's condition, whether clinical or experimental. Phrenologists often knew the character features of an individual being examined or conversely, gave interpretations vague enough to fit within the individual's own character assessment—much as astrology readings do. There was a tendency to fit the data to the preconceived notions of a particular brain organ.[65] Likewise, because of the practicalities of running fMRI studies, researchers may not be able to assign conditions or even the order of conditions randomly. The subtraction method used helps but does not eliminate the vulnerability to experimenter bias. If suitably primed, people see what they expect to see and so knowing either the clinical or experimental condition can shape and even influence the outcome. Inattentional blindness, for example, was stunningly revealed in a study of radiologists, 83 percent of whom missed the insertion of a gorilla into an MRI scan.[66] While this study showed how expectations lead even experts to miss a critical stimulus embedded in a scan, the converse in also true. We tend to want to confirm our hypotheses (in fact, it's a well-established social psychological concept of confirmation bias) and will behave in subtle, unconscious ways to help along that confirmation.

The interpretation of a phrenological reading of a cranium and an fMRI scan relies on experts who translate the data into the neurological or psychological narrative that is consistent with current theory. The participant whose brain is being observed is dependent upon the skill and specialized knowledge of the interpreter. Both technologies are indirect measures of brain activity and outward appearance, whether evidenced by cranial enlargements or colored areas on the scan that represent increased metabolic activity. Both depend heavily on researcher decisions of regions of interest and threshold points for establishing differences. Again, it is important to realize that Gall never meant to equate phrenology with physiognomy (e.g., Lavater's theory) where outward physical appearance equated to character.[67] It was the underlying physiology of the brain made visible through a faithful cast of the skull.

Likewise, both approaches attempt to be an objective measure of underlying activity. Imaging, in particular, is an extension of the idea of "medical diagnosis at a distance," a concept that began with the invention of the stethoscope—an instrument that initially sparked outrage among physicians at the time of its introduction in 1816.[68] Medical technology has historically had for its premise "a belief that objective medical findings can fully answer the question of what is wrong with patients."[69] These technologies served to extend the senses.[70] In the case of the stethoscope, the sense was hearing; one reason for its ultimately successful adoption was that it "provided access to a new range of physical events within the body that produced significant knowledge, made authentic and dependable by being detected and evaluated directly by the doctors themselves."[71] For the main technologies under discussion here, the sense was vision but it is easy to see the parallel. Perhaps it is instructive to note that another reason for the stethoscope's adoption, according to Reiser, was because the use of the instrument allowed examination with a barrier between the physician and the patient.[72]

In this regard, phrenology and fMRI are linked in being visual indices of underlying brain function and that visual depiction is presumed to have truth. Obviously, this visual feature is more compelling in the neural images but the ubiquitous phrenology heads with the faculties mapped on the skull served a similar purpose. George Combe's 1828 *The Constitution of Man*, that bible of phrenology for the next century, is replete with visual imagery and allusions: "clearer views," "seen," "observations," "enlightened," "illuminated," "shews [*sic*]," "reflection," all in just the first few pages of the section outlining phrenology.[73] Visual metaphors abound in the fMRI studies: images "reveal," "pinpoint," "illuminate," "illustrate," and "snapshot."

The fascination and now prolific use of current imaging technologies also has a direct lineage from the x-ray, another visual technology that puts the patient at a distance. In its early days of medical usage in the 1940s, there was debate about the relative superiority of the stethoscope versus x-rays in diagnosis. The x-ray prevailed in large part because of its relation to the superior sense of vision.[74] It was "viewed as the most dramatic extension of the sense into the body's interior ... the equivalent of an autopsy."[75] Reiser quotes a physician George Dock in 1921: "The dream was, and still is, to endeavor to detect and treat the disease instead of treating the patient."[76] This sentiment still seems to prevail today and while my discussion about phrenology and imaging is as much about their respective roles in elucidating the functioning the normal brain, both are also invested in elucidating the pathological brain. For example, Gall was convinced that mental illness was caused by brain lesions and thus evidenced on the cranium.[77]

One of the issues at heart is the "myth of photographic truth," the idea that images passively reflect reality rather than being constructed by individuals and influenced by their socio-cultural assumptions and backgrounds.[78] Visual information is accorded greater power among the senses and the imaging is perceived to be objective in their representations of underlying physiology. An old adage is "the camera cannot lie" and this belief is extended to all imaging technologies. Golinski suggests the initial impact of the photograph was precisely its creation by a machine—not the photographer—and that it seemed devoid of human agency.[79] The image was assumed to be literally a "re-presentation" of the thing itself.[80] This assumption, still carried on today, ignores the fact that images are constructed and this is even more the case with fMRIs. It is not uncommon for physicians to say things along the lines of "The MRI confirms the diagnosis or presence of such and such," thus reinforcing this jettisoned agency. Those high status images, produced by sophisticated machines whose workings are fully understood by relatively few, are thought to be unmediated slices of the brain—a view into the interior of the mind. And yet, this is totally contradictory to the fact that the image is a "constructed artifact."[81] "Scientists creatively select, adapt, and appropriate information and ideas from a range of possibilities to produce new technologies. The decisions they make in the course of producing these technologies, especially in the case of medical imaging, are intimately connected to the larger cultural context of visualization."[82]

The assumption of direct access to the physiology underneath the skull also derives from the "cult of precision" that accompanies modern science. That scientific measurement should be accomplished with instruments and as precisely as physically possible is a concept taken wholly for granted by the scientific community.[83] In the history of psychology this assumption is of special note because of the interaction with the observer and the typical subject matter, human behavior. The early sins of philosophical speculation, introspection, and subjective measures were rampant in psychology's history as the field developed into an experimental discipline. Having calipers with intricate demarcations for measuring the cranial organs in phrenology or having the arbitrary but now standard voxel to assess areas of brain activity represent attempts to convey scientific precision. "But the mere fact that a science measures, counts, and calculates does not make it an exact science."[84] Furthermore, the more precise the measure, the greater the individual differences appear for the variable under investigation. The results of both phrenology and fMRI represent "meta-individuals"; the famous phrenological head with its lines drawn to correspond to the underlying faculties or the scan with its varying colors to depict oxygenation are based on averages and, in the case of the latter, subjected to a great deal of research manipulation. "Precision has assumed the transparency of the obvious," yet that transparency is illusory.[85] The path to acceptance for a particular technology becomes obscured over time and what is lost are the incremental changes that were negotiated according to culturally accepted principles at the time. To be endowed with scientific authority, the culture determines what constitutes new knowledge.[86]

What can be learned about peering into the mind, whether by phrenology or imaging? The key take-away messages are as follows: (1) There is still a focus on localizing function in the brain. This is a useful position for much of our current understanding of the brain. However, without recognizing that localization reflects a particular perspective and that other perspectives, such as distribution and connectivity, might have some validity, we run the risk of losing valuable insights into the workings of the brain. (2) The images today are no more literal representations of brain areas than the markings delineated on a phrenological head. Functional MRIs are constructed analogies of a particular metabolic process that is only an indirect reflection of actual activity. (3) Visual data have enormous power in influencing interpretation of data. (4) Neuroscience, with its main manifestation of fMRIs, has the potential of influencing public understanding of human behavior in ways that exceed the actual data.

Notably in 2008, Weisberg, Keil, Goodstein, Rawson, and Gray performed a study wherein participants considered "naïve" in their scientific understanding versus experts were given verbal explanations of psychological phenomena.[87] A neuroscience finding purporting to illustrate the results but in fact, irrelevant, was presented in the verbal report for one condition. Importantly, the verbal report was either vague and circular—in other words, the results were logically flawed and thus a bad explanation—or presented a good explanation of the phenomena. In the explanation-only group, the flaw was detected by most participants; notably, the flaws were overlooked by the non-experts in the verbal report plus neuroscience condition. Fernandez-Duque, Evans, Christian and Hodges more recently performed a similar study, this time including an fMRI scan as a superfluous addition to the good or bad explanations.[88] Similar to the results of the previous study, they found that neuroscience additions influenced participants' judgments such that they believed the explanations, whether good or bad, were satisfactory. However, the addition of a brain scan did not yield any enhancement of the neuroscience effect. The researchers did not have a condition of superfluous image versus superfluous text to test the power of the visual stimulus in influencing judgments. Their results also showed that neuroscience explanations, even when superfluous, were privileged over social science explanations or even other physical sciences. The authors' conclusion was that neuroscience because of its focus on the brain, is the favored science for explanations of cognition and behavior.

My intention in this chapter is not to disregard the advances in our current understanding of brain function based on contemporary technologies and methodologies. Instead it is a cautionary tale of blind acceptance of the technologies without examining the underlying assumptions and theory. The wild enthusiasm from an uncritical public encouraged phrenology for decades longer than what was supported by the scientific community; we have a similar enthusiasm from the popular press, much of which depends on visual information for its audiences. As new technologies are being developed and utilized in research, such as diffusion tensor imaging that focuses more on connectivity rather than activity in a single brain structure, we may well see that some of the early and more extreme claims from fMRI technology will be relegated to the realm of pseudoscience themselves.

2. The Double Brain

WHILE THE FACT THAT we have pairs of spinal nerves emerging from the spinal cord that innervate each side of the body makes good anatomical sense, one of the most remarkable features of the human brain is that it, too, has two halves. We commonly refer to brain structures in the singular yet virtually all of them exist with mostly bilateral symmetry on each side. The two halves of the cerebral cortex, or "cap" of the brain, are the cerebral hemispheres. Even more remarkable is that each hemisphere is responsible for both controlling motor responses and processing sensory information from the opposite side of the body. In other words, the left hemisphere controls the right side of the body and vice versa. The two hemispheres are connected by two commissures, thick bands of myelinated fibers: the small anterior commissure and the much larger, prominent one called the corpus callosum.

Ground-breaking research on the organization of the cerebral hemispheres by Roger Sperry in the 1950s and '60s ultimately yielded him a Nobel Prize. What was so important about his work to merit that prestigious award? The simple answer is discovering that the two hemispheres were functionally asymmetrical and capable, under certain conditions, of operating independently: the "split-brain" phenomenon. The better answer was because of his contributions to elucidating how consciousness manifests itself in the brain—the ultimate psycho-physical isomorphism or mind-body problem.

His research studies were meticulous and well-designed; the published results on a sample of epileptic patients whose corpus callosum had been severed were captivating to both the academic community and general public.[1] Using a testing paradigm now outlined in every introductory psychology and neuroscience textbook, they were able to present information just to one hemisphere at a time. Researchers were able to demonstrate in these original patients and subsequently others across numerous studies that the two hemispheres were characterized by distinctly different abilities of various cognitive functioning. Most notably, the left hemisphere or "left brain" was verbal and the right hemisphere or "right brain" was more nonverbal and associated with spatial perception and object recognition. Today neuroscience textbooks commonly proclaim that the left hemisphere is responsible for language and linear (propositional) logical thought processes, while the right is responsible for spatial relations, face detection, and object recognition. Less accurately, general discourse and the popular press refer to the right brain as the more creative, musical, and mathematical hemisphere. The grain of truth in the latter statement is contained in findings of studies that do show the right hemisphere exhibiting more activity during tasks related to art, music and geometry. These are all abilities/tasks that involve spatial relations but it would be simplistic to assign large categories of abilities to each hemisphere. For example, the right hemisphere is not entirely nonverbal and the left hemisphere functioning is actually superior to that of the right hemisphere in performing arithmetic tasks. Likewise, while the right hemisphere plays a major role in the perception of music by the brain, different elements of music are processed in each hemisphere.[2] More accurate would be Sperry's contention that the left hemisphere is more analytical in function and the right hemisphere operates more holistically and perhaps is involved more in emotional processing.[3]

The work on split brains has generated much fascination among the public and given rise to applications in a variety of domains, ranging from education to business to criminology. Before long the hemispheric specializations were simplified, type-cast, and value-laden. Harris points out that lateralization of the hemispheres was included in a well-known list of concepts that constituted cultural literacy for Americans in the 1980s.[4] Corballis mentions that in 1977, the popular magazine *Psychology Today* named the split-brain phenomenon the "fad of the year," reflecting the enormous interest in the topic.[5] The left hemisphere was the logical brain in contrast to its right, emotional opposite. Since then teachers have been admonished to present information in ways to reach both hemispheres and children are encouraged to enhance their creativity by developing their right brains (c.f. *Drawing on the Right Side of the Brain*, now in its fourth edition with over three million copies sold, according to Amazon).[6] Never mind the fact that most of us have never had a callosotomy and have fully integrated hemispheres and unlikely to have stimulus presentation apparatus that can differentially apply information to an individual hemisphere.

Sperry's graduate student, Michael Gazzaniga, continued with an academic career of research of his own on hemispheric specialization, becoming better known today than his mentor. Recently, he has reflected on his life's work and synthesized many decades of research on the split brain.[7] Gazzaniga proposes that the most important function of the left hemisphere is to act as "narrator" or interpreter. This interpreting function ultimately gives rise to the integration of consciousness and provides a sense of meaning through a (sometimes) concocted narrative.[8] The specialization of each hemisphere, Gazzaniga maintains, is the result of evolutionary selection over time. The role of the brain is "to make decisions that enhance reproductive success."[9] The successful adaptations of the brain have resulted in hemispheres that operate more efficiently due to decreased redundancy and increased "work space" for neural circuits. The "real estate" of cortical functioning has been expanded by such specializations and the consequence is a double brain.

But while the term "double brain" has often been utilized, it is a bit of a misnomer. Some functions are truly duplicated by each hemisphere. This is true of basic perceptual, sensory, and motor processes up to a point.[10] However, these functions operate independently of one another. This is seen with one of the astonishing tasks performed by split-brain patients that involves simultaneously drawing different figures with both hands with apparent ease.[11] Luck, Hillyard, Mangun, and Gazzaniga demonstrated that the two hemispheres can process attention quite independently and in parallel.[12] Despite the long-standing principle that for most individuals, the language center is housed in the left hemisphere, experiments with the split-brain patients reveal a more nuanced version. For example, patient studies reveal that spoken and written language can be processed by different hemispheres. Likewise, the right hemisphere, while superior to the left in visuo-spatial tasks, can be matched by the left hemisphere in certain visual tasks without the spatial element.[13]

The right hemisphere has perhaps undergone the most significant "rebranding." Once silent, supposedly only good for some spatial recognition, it has achieved the desired status of being the creative and mysterious side of the brain. Conversely, right hemispheric dominance has connoted masculinity, especially with regard to mathematical ability. The male sex hormone testosterone was hypothesized to retard the growth of the left hemisphere.[14] There is a certain irony when creative writing programs implore their participants to "access" the non-verbal hemisphere to improve their writing![15] Then, just as the left hemisphere acted as the language "interpreter" for the severed right, the right hemisphere was found to be the perceptual "interpreter" for the severed left.[16]

When publications on the research findings of studies with split-brain patients erupted in '60s and '70s, there was an atmosphere of new discovery and ground-breaking work. Some of the hype was certainly justified—never had there been such systematic study under controlled laboratory conditions of a sample of individuals who had undergone a callosotomy. Renewed interest (if indeed there has been any waning of such in the intervening years) has been sparked by the utilization of newer imaging technology, both to confirm the complete severance of the corpus callosum and anterior commissure in the remaining sample and to study the differential hemispheric activation from various cognitive, emotional, and sensory tasks.[17]

However, as with much of the field of neuroscience, the current understanding of the split- or dual-brain tends to be ahistorical. Yet the topic enjoys a relatively long period of punctuated discussion especially during the localization versus equipotentiality debates during the 19th century. The dual brain arises as an important element in understanding the biological basis of consciousness—a topic to be discussed in more detail in the epilogue—as well as the search for locations of particular brain functions. Indeed, the discussion is intertwined with many other issues examined in this book and so there is some inevitable redundancy in tracing the development.

The fact that humans have two "halves" of the brain is obvious to anyone who has physically inspected a brain. It is not surprising that the writings of physicians in ancient Greece report this finding and associated different brain functions with various regions of the brain.[18] They were certainly aware of the contralateral control of hemispheres with their respective bodily halves. It was more common, however, to view that localization longitudinally through the brain rather than laterally. In other words, the two halves were mostly presumed to be identical in functions which were organized by the brain's longitudinal axis. Perception was thought to be situated in the anterior portion of the brain, reasoning in the central, and memory in the posterior portion. The ventricles were the structures (or more accurately, cavities) of interest for these functions rather than the brain tissue itself.

The four chambered ventricles are placed along the longitudinal axis and circulate cerebral spinal fluid within the central nervous system.[19] The fluid is secreted from the choroid plexus that lines the ventricles and ultimately exits the fourth ventricle via the central canal into the center of the spinal cord. Small openings also allow fluid to circulate around the exterior of the brain, eventually being reabsorbed in the cerebral veins. The cerebrospinal fluid both nourishes and buffers the brain within the hard bone of skull.

The interest for the ventricles for the Greeks, however, was because "the pneumatic spaces seemed to be more appropriate for the non-corporeal soul to interact with the body than ... the substance of the brain itself."[20] This ventricular model persisted for many centuries, perhaps more for theological reasons as it did not present a challenge to the idea of an immaterial soul. Medical works in the 1500s, such as Gregor Reich's *Margarita Philosophica,* discuss the communication of sensation, perception and imagery in the anterior or lateral ventricles with the cognitive processes of the third ventricle, which are then communicated to the fourth ventricle to produce a memory. This was, in fact, a rendition of the modern reflex arc. Descartes' position in the 17th century was also to emphasize the role of "animal spirits" (versus *pneuma* of the Greeks) but now as well to shift focus on the nerve fibers rather than the ventricles. Nonetheless, laterality of the brain was being viewed as unitary in function.

One opposing view has been documented in a 12th-century medieval text that referenced an anonymous classical Greek medical text from the fourth century.[21] Attributed to a follower of Diocles, the relevant passage signals what is perhaps the earliest description of a "dual brain." The passage, translated from the Latin, reads as follows: "Accordingly, there are two brains in the head, one which gives understanding, and another which provides sense perception. That is to say, the one which is lying on the right side is the one that perceives; with the left one, however, we understand."[22] Diocles believed the brain was a "second soul" after the heart and disruption of the flow of psyche between the heart and the brain was the source of mental illness. However, his theory of cerebral lateralization did not catch on in mainstream Greek medicine and so was largely ignored in the following centuries. It is interesting, however, that the distinction of functions of reason versus perception between the right and left hemisphere is later echoed in the 19th and 20th centuries.

Harrington posits that the explanations for the double brain during the Renaissance years were attempts at reconciling a theological problem rather than a medical mystery.[23] In order to maintain the dualistic view of a separate soul and body, there needed to be some point of union or intersection of their actions. Descartes offered the pineal gland at the base of the brain as the structural manifestation of this union because it is centrally located and singular in its make-up, unlike the duplicated structures in the cerebral hemisphere. This hypothesis was ultimately not supported but the idea of a *sensorium commune* persisted in investigative physiological and medical studies.

Franz Joseph Gall, the founder of phrenology, largely gets credit for establishing once and for all that the brain rather than the heart was the organ responsible for mental processing and behavior.[24] But at the turn of the 19th century, the theological implications of a material mind were still a concern. This concern, in fact, led to Gall's departure from his home country of Austria to seek the more favorable intellectual climates of France and Britain. Even so, he tried to placate religious authorities by saying that while the faculties of the mind might be housed within the brain, the fact that humans possessed a faculty to know God was sufficient evidence for the divine control over the body.

The fact that the brain was double in form meant to Gall that we had exact duplicate mental faculties in each hemisphere—truly a double brain. Theoretically, each hemisphere was capable of functioning independently. However, the duplication was in the interest of redundancy. One hemisphere lay in reserve while the other expended its efforts rather than operating simultaneously. This redundancy likewise offered a protection against brain injury. Damage to one part of the brain did not necessarily result in total loss of function because the intact, opposing structure could take on the duties of the stricken one.[25] When evidence from clinical cases showed that in some instances unilateral damage did indeed result in loss of total function, the phrenology response was to suggest that this was because the normal "balance" between the hemispheres had been upset. This response is representative of one of the fatal flaws of phrenology—an answer for everything.

Despite the fact that close inspection shows the gyri and sulci of each hemisphere are not, in fact, perfectly symmetrical, when Gall was developing his theory of organology in the late 18th and early 19th centuries, hemispheric tissue from an anatomical standpoint was thought to be uniform and have no distinctive patterns. The symmetry was required so that the brain could operate the two halves of the body in an organized, cooperative, and seamless fashion. If the two hemispheres were not identical in function, there could be bodily confusion, lack of integration, and perhaps most importantly, a lack of a unified impression or consciousness.[26] Renowned French anatomist Marie-Francois-Xavier Bichat proposed that a brain was composed of "an ensemble of functions that collectively offered a resistance to death."[27] This anatomical substrate was then superimposed over a unitary mind (otherwise known as the soul). This was "Bichat's law of symmetry."[28]

Dysfunction of brain could indeed influence dysfunction of mental faculties. Could there be instances where the dual brain did not act harmoniously or in synchrony? Benjamin Rush, physician and signer of the American Declaration of Independence, certainly thought so and gave the example of sleep-walking to illustrate dual minds in conflict.[29] Similarly, in 1836 Hewitt Watson said that dreaming showed evidence of this dual mind because in a dream one could have impressions of (and even dialogues between) both oneself and others.[30] Cases of dual personality might also be explained by a dual brain.[31] Queen Victoria's physician, Sir Henry Holland also agreed, indicating that pathologies such as somnambulism and hysteria were examples of the dual brain at work.[32] Uncharacteristically for the time but more prescient than most of his contemporaries, he attributed coordinated movement and unity of sensory-perceptual experience to the connecting band of fibers, the corpus callosum. The corpus callosum then was typically ignored by scientists, who believed it provided only structural support in the brain.

The most prominent voice on the topic prior to the later major localization debates of the 1860s was another British physician, Arthur Wigan, author of *A New View of Insanity: The Duality of the Mind.*[33] Wigan and Holland were the two medical men advocating a dual brain theory but Wigan thought Holland had not gone far enough and proposed a theory with wide-ranging applications to mental health, education, and criminology. Wigan believed adamantly "*that the mind is dual, like the organs through which it is manifested.*"[34] He turns to the two eyes as an analogy to understand the two halves of the brain. Each eye is separate, capable of operating independently, but under normal conditions works together with the other for a unified visual experience. He dismissed the corpus callosum as having any pertinent role other than to provide mechanical, structural support between the hemispheres.[35] Instead, there was "the great central nucleus" by which either brain could affect both sides of the body, one hemisphere doing so directly and the other by "consentaneous action."[36]

Wigan's theory contained 20 propositions, which he discussed briefly in his medical journal article and expanded upon in his larger manuscript.[37] Relating to normal brain functioning, each hemisphere was a "distinct and perfect whole, as an organ of thought"; separate thought processes can occur simultaneously in each hemisphere; each hemisphere is capable of separate (and often opposing) volition; and one hemisphere is "superior in power to the other" and can activate or, conversely, inhibit the other.[38] Clarke notes this latter proposition contradicts his central premise of equal, dual brains.[39] However, resorting to a superior hemisphere is not the typical mode of operation for Wigan's brain. Wigan also divided the hemispheres into three lobes (not the four as is customary today), each having its own faculty.[40] Wigan came down squarely on the side of localization, positing that different parts of the brain are responsible for different functions. There are duplicate brains, which ensures redundancy and protection against disease or injury. He maintained mind/body dualism, however, noting that "without the brain to connect *mind* and *the material world* perfectly, he must be either a mere animal of instinct, a madman, or an idiot."[41] The mind operates through the medium of the brain.

When the two hemispheres do not act in unison or rather, are in conflict, mental pathologies ensue.[42] Anxiety, for example, occurred as a result of asynchrony between the hemispheres. Winslow reporting in 1849 on an unpublished manuscript of Wigan's after the latter's death, states: "The distress brought on by this inability to guide the thoughts—a frequent consequence of great anxiety—this inability to use the two brains concurrently; that is, to exercise *attention* or study, is one of the most pitiable states of mind that can be conceived.... In following the ideas of another man he can generally leave his own intellectual organs in quiet; the discordant action of the two brains may thus subside perhaps into repose, and on resuming their duties they may have re-established the unison and consentaneity which is necessary to the tranquil exercise of the mind."[43]

Wigan believed that insanity was due to compression of one of the hemisphere leading to conflict between the two minds. In cases of dementia near death, "one brain always 'goes out' before the other; but previous to its extinction in this gradual manner, it may obey the commands of its more energetic brother when thoroughly roused, long enough to dictate a will which may save a family from destruction."[44] This idea gets at some of Wigan's other propositions from the list of 20. In the case of a functional (versus organic) brain disease, the healthy hemisphere can still exert control up to a point over the diseased one but the degree of control depends on the severity of the disease and the ability of the patient to exert self-control.[45] The latter concept was a key component of Wigan's theory as he believed that through education, one could learn to control voluntarily a hemisphere's actions.[46] Lack of control resulted in various forms of insanity, the mildest of which was the presence of delusions. In insanity, one can "trace the admixture of two synchronous trains of thought, and that it is the irregularly-alternate utterance of portions of these two trains of thought which constitutes incoherence."[47] The two trains of thought might be composed of one that is rational and one irrational but regardless, the end result is one of conflict within the self.

The latter decades of the 18th century and first half of the 19th century were a period of intense scientific interest and exploration of topics that today may smack of pseudoscience. Besides phrenology, mesmerism (hypnosis), dual consciousness or personality (now referred to as dissociative states), spiritualism, somnambulism (sleep-walking), hallucinations, déjà vu, and various forms of insanity all were under medical scrutiny. Within this environment, Wigan and Holland's ideas were received with some interest. Ostensibly, having two independent brains could account for these phenomena. Wigan's explanation for déjà vu, for example, was that one hemisphere was "asleep," while the other tried to catch up with it.[48] Holland believed dual consciousness or personality was caused by alternating hemispheric activity.[49] Double consciousness occurred "where the mind passes by alternation from one state to another, each having the perception of external impressions and appropriate trains of thought, but not linked together by the ordinary gradations, or by mutual memory."[50] However, Clarke states that overall, the reception to Wigan's views in particular was tepid.[51] Phrenologists argued that there was no originality in the idea of separate faculties and organs—indeed, that was a fundamental premise of the field. But more problematic was simply the lack of evidence beyond conjecture. The standards of what constituted scientific proof were not met, not even with a collection of clinical cases as was commonplace in medical writing at that time. The basic ideas were plausible to his contemporaries yet the theory lacked organization and persuasive evidence. Holland's ideas perhaps fared a little better because he was much more cautious in his conclusions and applications.

The notion of the dual brain would lay quiescent for a generation until being revived in the late Victorian era. Very few questioned the symmetry of the brain, structurally or functionally. There were a few anatomists, such as Vicq d'Azyr, who noted early in the 19th century that the outer surface of the two hemispheres in humans were not precisely identical. But rather than proposing any consequent functional difference, he took the variation as a sign of superiority over lower animals. Functional differences between male and female brains were also noted but only the occasional voice suggested that these differences corresponded to lateralization differences.[52]

The next key developments in establishing lateralization of function arose from the debate among medical men in France, long a country renowned for its contributions to the fields of anatomy, physiology and the emerging field of neurology. Gall had relocated to Paris upon leaving his Vienna home in order to promote his theory of phrenology unfettered. He was to be soon disappointed, however, as the medical and scientific community mostly rejected and then actively campaigned against Gall's views. The French accepted the idea of separate faculties of the mind but not the underlying organology.[53] Gall's scientific nemesis was Marie Jean Pierre Flourens, whose experimental work in the early decades of the 19th century convinced him that the brain acted as one and equipotentiality characterized the substance of the brain. So again, a fundamental principle of brain action is debated: localization versus distribution or equipotentiality. Bichat's "law of symmetry" held firm.[54]

This debate reached its Victorian pinnacle in mid-century as scientists considered both clinical and experimental evidence in the emerging field of neurology. Harrington refers to this 20-year period (1840s–1850s) as a time of a paradigm shift.[55] The mechanistic principles that had dominated the scientific landscape from the classical era were giving way to energy dynamics, nerve cells, and reflex action. Physician Thomas Laycock proposed the brain was no different than the other nervous system ganglia that had been identified at the spinal cord level.[56] In other words, the brain was a ganglion in a chain of ganglia that made up the central nervous system.

When Paul Broca presented his clinical cases beginning in 1861, including the case of the now famed individual referred to as "Tan" (because it was one of the patient's few intelligible sounds), the evidence seemed to point to localization of the brain for language. Even more importantly, because Gall's phrenology had already included an organ of language situated bilaterally in the brain, Broca's work suggested that it was the frontal lobe of the left hemisphere that was responsible. This finding appeared to contradict Bichat's law of symmetry but supported the idea of localization proposed by Bouillaud.[57] Jean-Baptiste Bouillaud had suggested that the frontal lobes housed both the representational functions of language, as well as the ability to articulate language. He speculated that the two functions were distinguished by the respective white and gray matter of the frontal lobes—a now obvious incorrect belief. Besides the unilaterality of the brain damage and the focus of the frontal lobe, Broca's location also did not align with the specific area that Gall had proposed for the faculty of language. Interestingly, Broca initially was unwilling to concede a natural left hemispheric dominance for language but rather viewed it as a function of learned behavior.[58]

Further controversy erupted when Gustav Dax, the son of French physician Marc Dax, came forth to claim that his father had made and reported the same clinical discovery of left frontal lesions being associated with language deficits in 1836, some 25 years earlier! Bouillaud and others in the Academy of Medicine sat on the information until Dax the younger decided to publish the 1836 manuscript along with supporting information. It should be noted, however, that no proof of presentation of the material in 1836 to a scientific or medical body has been found.[59] In contrast, there is evidence that Broca's work was based on clinical observations recorded long before the revelation of Dax's manuscript.

This battle of academic precedence consumed the Academy for several months. But for Broca, it was more than just getting credit for the idea of localization of speech. Dax's views challenged the status quo, namely, Bichat's law of symmetry. One of the most fundamental principles of nervous system functioning was potentially being questioned. And so more evidence to build his case was necessary.

While Broca's first public remarks in 1861 were tentative in regard to left hemispheric dominance for language, his 1865 paper was more direct. The year 1861 in France was still characterized by the belief that both cerebral hemispheres were considered equal halves. Four years later Broca had assembled more additional evidence with which to present a contrasting view.[60] Dax's work had indicated a left frontal lobe location for speech articulation and this was confirmed by Broca's clinical cases with subsequent anatomical verification provided by autopsies. By the end of the decade, Broca's stance on lateralization for speech was largely accepted by the medico-scientific communities.[61]

But even then, there were mixed views as to whether the left-right cerebral differences were inherent or learned. French anatomist Pierre Gratiolet was able to show that the anatomy of the two frontal lobes was, in fact, not identical, with the left frontal lobe typically larger than the right.[62] Additionally, his work on fetal development showed this difference emerged early prenatally. Broca still tried to reconcile Bichat's law of symmetry with the growing body of evidence that the hemispheres were not identical with respect to language. He did so by proposing that there were no innate such differences but rather the differences were due to an "artifact of the developmental differences between the two frontal lobes.... The two lobes began life with identical intellectual potential, but in simple terms of physical growth, the left was the more precocious side ... *thus we educated it to the exclusion of the other.*"[63] Similar to earlier Victorian views, in some cases the right hemisphere was thought to substitute for the left and responsible for recovery of various intellectual faculties. Given the new developments in histology mid-century, this activity could happen at the nervous cell level and thus go undetected with inspection of the gross anatomy of the brain. For Broca, this helped explain the conundrum of having two "brains"—now clearly *not* identical—yet having a unified experience. The emphasis on education reflected human's unique facility for language, something that lower animals were not thought to have and thus reflected in the symmetrical appearance of their brains.[64] It was a short leap in the post–Darwinian years of the Victorian mid-century to the originally neutral notion of the superiority of hemispheric asymmetry proposed by Broca being used as a justification to assert social superiority, whether by race, class or gender.

Once the door had been opened by Broca to accept asymmetry of the brain with respect to speech articulation and aphasias, localization of other intellectual faculties began to follow. Neurologist Carl Wernicke introduced data to support differentiating a receptive aphasia from the expressive or Broca's aphasia and indicated a specific location in the left temporal lobe rather than the frontal lobe. Indeed "all sorts of cerebral centers for different mental functions (reading, writing, calculation, abstract conceptualization) were happily identified and incorporated in one or another variation on the Wernicke model" by neurologists during the 1870s.[65] Even the psychological faculty of "Will" was thought to arise from the left, as evidenced by apraxia, an inability to touch or grasp a designated object. The left hemisphere was clearly involved with purposive movement and Hugo Liepmann proposed that aphasias and apraxia were linked by their similarity in the inability to carry out certain motor responses.[66]

What could be the physical basis underlying this left hemispheric superiority? There was still an underlying belief among scientists of an "unequal but *compensatory* anatomical differences."[67] Blood flow was a prime candidate for the responsible agent but that hypothesis was soon disproved when anatomists were able to show that the blood supply to each hemisphere is shared. Furthermore, there were no differences between the left and right carotid arteries as would be expected by differential blood flow.[68] Broca's explanation was that there was a "growth gradient" whereby "desynchronous growth of the 'motor' and 'intellectual' parts of the brain could lead to a dissociation, both within and between the hemispheres for control of speech and handedness."[69] Brain weight and density were also hypothesized as causal mechanisms but there were mixed results. Scientists did contortionist acts to rationalize contrary results to fit with their particular existing views.

In the meantime, while not subjected to the same kind of attention from scientists as the left, the right hemisphere was gradually coming into focus. At first considered the "minor" hemisphere, the right was viewed as subservient to the left—a silent, weaker partner.[70] John Hughlings Jackson, eminent British neurologist presented a different view of the right hemisphere. If the "faculty of expression resides in one hemisphere, there is no absurdity in raising the question as to whether perception—its corresponding opposite—may not be seated in the other."[71] His clinical observations led him to the conclusion that recognition of objects, places, and persons were the provenance of the right hemisphere. Damage caused an "imperceptions" or agnosia.[72] However, he still considered the left hemisphere to be the "leading" half of the brain for expression, while the right hemispheric action was the "leading side for perception—educated senses" and its actions were "involuntary or automatic."[73] Subsequent neurologists would refine Jackson's initial ideas to focus on spatial perception as a distinct function of the right hemisphere.

In fact, several neurologists had noted that the right hemisphere was involved at a more basic level (e.g., emotional) in the faculty of language. As Benton points out, the differences in language function between the left and right hemispheres are relative, not absolute.[74] Even Broca's famous patient Tan made utterances and the occasional oath when disturbed. Jackson was cautious, however, in equating localization of symptoms to localization of function. Language was a "dynamic process deriving from the integrated functioning of the entire brain, such that the more complex the task, the more brain regions involved, with each making its own contribution."[75] As neurologists began to accrue evidence more systematically from their case studies, it became apparent that the right hemisphere indeed had a significant role in emotion. In support of this view, some (e.g., Jules Luys in 1881) observed that left-hemiplegic patients exhibited more emotionality than did right hemiplegics. More controversially, somatic symptoms of hysteria were also thought by some to be more prevalent on the left side of the body. This latter notion engendered some debate in the last decades of the 19th century. Even among those who agreed with the findings questioned whether the laterality was more of an artifact of some other physiological abnormality at work in hysteria. In 1899 Pierre Janet reported a study of lateralized symptoms among hysterical patients. His findings revealed mixed results—there was an increased tendency for language-related symptoms with right-sided hysteria, while the left side of the body was correlated with symptoms of somnambulism, dissociation, and seizures. For Janet, this confirmed that the right hemisphere was "supersensitive."[76] Luys thought the right hemisphere was likewise larger among the insane—not because of right brain over-development but rather due to left brain atrophy.

These observations and studies by others contributed to a shift in medical parlance towards viewing the right as the irrational half versus the rational left hemisphere. It was not long before the irrational hemisphere was considered to be the dominant hemisphere among females, while the rational brain was, of course, the masculine.[77] This view had support in that it was commonly believed during this time that females were dominated by the posterior portion of the brain, the occipital lobe. Since this area was more developed on the right hemisphere, that gave further credence to the link between female and the right hemisphere.[78]

Extrapolations to racial differences also emerged. In the 1870s French physician Gaétan Delauney proposed that the superiority of the left hemisphere emerged with the "male element" during embryonic development and was further manifested in the superior races.[79] Evidence included the tendency to walk and turn to right among Caucasians, while the "middle inferior" races, such as Asians and Latinos tended to the left. The most inferior races had such poor brain differentiation, they showed no inclination one way or the other![80] Similar lateralization differences were seen among the various social classes. The higher classes had more developed left hemispheres (which could then wear out from too much intellectual activity, as evidenced by the increased frequency of neurasthenia). Soon hemispheric differences were conceived both in terms of dichotomous categories and rank order. Brain asymmetry was explained in such a way as to promote and justify the social system status quo.

It seems inevitable that handedness would emerge from these years of clinical debate on hemispheric dominance. Prior to the 1860s, handedness was not remarked upon or even notated in neurological case studies. Broca linked handedness with "brainedness" based on Gratiolet's evidence of asymmetrical fetal brain development.[81] The left frontal lobe developed its gyri faster than the right and the right occipital lobe likewise developed faster than the left. This was thought to account for the predominance of right-handed individuals in the population. According to Broca, humans were born with a left hemispheric advantage, predisposed towards intellectual faculties, and that predisposition would be reinforced by education in a positive feedback loop. Many physicians and early developmental psychologists at the end of the century noted that infants did not exhibit handedness at birth and there were mixed results of respective hemispheric weight and symmetry among infant studies.[82]

The association of left-handedness and the emotional right hemisphere soon led some scientists to suggest a link with madness. For example, although not replicated in subsequent studies, Jules Luys in 1879 had reported that the right hemisphere was heavier in the insane. Ireland in 1880 noted that children were more likely to be left-handed when compared to adults (ignoring the fact that some children would be forced to use their right hand once they were in school) as were "imbeciles."[83] Lombroso, founder of the field of criminology, not only believed that there was a greater preponderance of left-handedness among female, children and "savages," he thought it distinguished criminal "lunatics" as well.[84] "As man advances in civilization and culture, he shows an always greater right-sidedness as compared to savages, the masculine in this way outnumbering the feminine and adults outnumbering children," he wrote in 1903.[85] He did allow how not all left-handers were criminals but that "united to many other traits, [left-handedness] may contribute to form one of the worst characters among the human species."[86] Likely these views by Lombroso and others followed centuries of cultural press to favor right-handedness, as approximately 90 percent of the population are right-handed.[87] Aristotle associated the right hand with "the straight, the good, masculine" and the left hand with "dark, curved, evil, and feminine."[88] There seemed to be an assumption that the bilateral organization of the body had to be expressed by dichotomous categories of polar opposites.

So, in the last decades of the 19th century, a shift of sorts occurred. The idea of dual brains had been quiescent for a generation even as asymmetrical functionality was beginning to be established. The view reemerged in the medico-scientific discourse as some physicians speculated again that certain abnormal behaviors and neurological dysfunction might be explained using that concept. If the two hemispheres were indeed asymmetrical in function as well as structure, then perhaps there was a case to be made for independent, dual action. In the 1860s, experimental physiologists had hypothesized that the conscious mind "was an emergent product of a myriad of unconscious processes, which in turn were dependent upon cerebral activities."[89] Dual hemispheres produced two separate streams of consciousness; the binding occurred perhaps in the commissures or in some deeper structure.

Harrington notes that the themes of a divided Self or consciousness permeate not only the medical community but the public at large.[90] Robert Louis Stevenson's novel *The Strange Case of Dr. Jekyll and Mr. Hyde*, published in 1886, very much reflects this scientific discourse that was widespread at the time. That fictional work depicted the dual personality that could exist with the dual brain. There was also keen interest in the technique of hypnosis (really the same mesmerism that was practiced in the first half of the 19th century) with the work of Charcot with hysteria patients in Paris. The spiritualism craze that had begun after the American Civil War provided a natural application for dual consciousness as well. One hemisphere might be in touch with the supernatural, while the other's consciousness was grounded in the earthly reality. Harrington notes that then as now, scientists were having to "fight the creeping tide of the occult."[91] However, scientists in the late Victorian era were far more likely to accept some of these notions as phenomena worth investigating and to put the premises of the supernatural to the scientific test. Spiritual mediums were thought to be over-represented by left-handers, indicating right hemispheric dominance. A medical case widely cited when published in 1895 concerned the action of dual brains.[92] The case concerned a patient with two "separate and distinct states of consciousness, and in whom the right and left brain alternately exert a preponderating influence over the motor functions."[93] The 47-year-old former sailor spoke Welsh and was left-handed while under the influence of one consciousness but spoke English and was right-handed when the other (and "normal") consciousness held sway. This clearly demonstrated to many, including the attending physician, the actions of two independent cerebral hemispheres. Likewise, automatic writing, struggles with obsessive thoughts, and periodic episodes of lucidity in an otherwise insane individual were all potentially explained by double consciousness. Under stressful circumstances or the predisposition from inherited tendencies, the normal unity of the mind could split and fall into "pathological independence or disequilibrium."[94]

The late Victorian medical discussions of dual consciousness, however, are best exemplified by the writings of British alienist and physician Henry Maudsley and the French (actually Mauritian) experimental physiologist and neurologist Charles Edouard Brown-Séquard. They were the most vocal champions resurrecting Arthur Wigan's mid-century views of the double brain. Brown-Séquard was the most extreme, proposing that each hemisphere is a complete brain, "endowed with all the powers that we know belong to the whole cerebrum."[95] He disputed asymmetry on the basis that nerve cells were all alike and therefore each nerve center in the hemispheres were capable of performing the functions of others, if necessary. One could have considerable damage to one hemisphere yet not lose functionality. When there was a loss, that was due to inhibition, the "suppressing influence exerted on all the nerve-cells that have a share in that function, in either the encephalon or spinal cord."[96] In a healthy brain, there was a "dynamical equilibrium" maintained by the two cerebral halves.[97]

In contrast, Maudsley believed the brain to be not a double organ structurally but one organ functionally with two similar halves.[98] He disagreed with Brown-Séquard's hardline non-localization stance at least in regard to function, noting the left hemisphere's particular role in language. "It [the brain] represents at the same time the halves of the body and the unity of the whole whereof itself is part."[99] The two halves could operate singly, alternately, but typically work together with corresponding functions, although Maudsley admitted that one might exert a greater influence due to education. Consciousness alternated from one hemisphere to the other rather than a single consciousness simultaneously arising from both.[100] The perception of a unified experience came from the hemisphere's being "infected by the same *desire* or sense of purpose welling up from the lower-order affective organs" producing a "unity of the organic life."[101] Unity came from below the level of the cerebral cortex, in other words. His evidence included the observation that it was impossible to think two thoughts at the same time; instead the thoughts were successive in nature.[102] Dreaming offered another example. When a person is asleep, the hemispheres, "lacking the force necessary to full unity of action and yet taking up the organic impressions from the body, as well as any chance-impressions from without, manufacture from them the most incongruous dream-images and events."[103] Maudsley believed that insanity in the form of what we would today call bipolar disorder (formerly manic-depression) also reflected abnormal alternations between the hemispheres. If one hemisphere were damaged, the other might not be able to compensate for the specialized function "because it has not been taught the fine special work."[104] Similarly, practice enables the pianist to play with both hands as the motor responses become routinized. When first learning how to play, entraining the one hemisphere to the other's will is a major challenge.

The focus on the double brain also was concomitant with a resurgence of interest in hypnosis, leading to some of the most interesting studies carried out in the final decades of the 19th century. If the two hemispheres were independent and yielded two minds, then physicians concluded that it must be possible to hypnotize one and not the other. In France, where neurologists such as Janet and Charcot were demonstrating success in treating hysteria using hypnosis, there was some thought that nervous energy could be transferred from one side of the body to the other. Charcot believed that hysteria was localized in the cortex, likely on the left, despite the absence of observable lesions.[105] Transference for therapeutic purposes could be augmented by magnets or even the placing of metal discs upon one side of the body as in the practice of metalloscopy.[106] The latter procedure had shown success in relief of hysterical hemi-anesthesia. Both practices had been utilized (and abandoned) by phrenologists almost a century prior. This time, however, the Société de Biologie in Paris was enlisted to validate the efficacy of these practices. While its investigation revealed short-term relief from the procedures, eventually physicians associated with the Nancy school of medicine (e.g., Hippolyte Bernheim) were able to show that just as in the case of hypnosis, efficacy was due to psychological suggestion and placebo effect. The hypnosis debate between the Nancy and Salpêtriére schools, as exemplified by Bernheim and Charcot, touched on issues related to the premise of two independent hemispheres.[107]

The notion of the double brain, however, had its critics even then.[108] While all individuals had "two brains," according to neurologist Pierre Janet, not all were "madmen, nor somnambulists nor mediums."[109] Furthermore, many of the phenomena offered as evidence of dual consciousness were functions of time (successive rather than simultaneous presentations) rather than space. Increasingly, attention focused on the role of the corpus callosum as the binding agent of consciousness. By 1900 with the publication of a case study of apraxia, German neurologist Hugo Liepmann was able to demonstrate conclusively that the corpus callosum was critical for hemispheric communication. This finding helped turn the tide away from the double brain as the 20th century dawned.

Finally, at the same time, the advent of Freud's psychoanalytic theory meant a paradigm shift, divorcing psychiatry from neurology and making a turn from the physiological to the psychological causes of mental disorders. The result was to undermine the double brain theory even though it had influenced his own thinking.[110] The irony is that Freud, himself a neurologist, spent several years prior to 1900 working in vain on a theory of the mind grounded in the neural and reflex physiology of the day. This *Project for a Scientific Psychology*, published posthumously, utilized ideas regarding conscious and unconscious ideation that stemmed from Hughlings Jackson's work on the two hemispheres.

In any case, whereas Victorian science had a more holistic approach to body and mind, as exemplified by the double brain, the emergence of psychoanalytic theory returned medicine and psychology to the familiar Cartesian dualism. There were once again diseases of the mind (formerly soul) and diseases of the body, each with their specialists. For example, hysteria, once thought to be caused by dysfunction of the brain and nerves, was now relegated solely to the province of the mind. Harrington refers to this shift as the "resurrection of Cartesianism in the clinic."[111]

However, at the same time, some aspects of holism remained in regard to the brain. While localization of certain functions, such as language, was widely accepted by neurologists and neurophysiologists, there was an overall swing back towards a distribution model. Karl Lashley, an American physiological psychologist, proposed in the 1920s his principle of mass action that had the notion of equipotentiality once again at its core.[112] He famously embarked on a scientific search of the engram, the biological memory trace. His experiments failed to show the localization that he expected to find, although we now know his methods were flawed. What was now important was the quantity, not the location, of the cortex that was affected in a lesion and the ability of the brain to recover function: neural plasticity. Scientists did not totally ignore localization related to the aphasia debates of the mid–19th century but those were largely relegated to general abilities now and not discrete functions.

Roger Sperry was a student of Lashley's and so his work was originally undertaken to find support for equipotentiality. Famously, his animal experiments showed the opposite. When the corpus callosum and optic chiasm was severed in order to have information going to only one side of the brain, results suggested independence between the two hemispheres. It was this research that led to the now famous callosotomy procedure in humans that gave rise to the so-called "split-brain" patients. Under special laboratory testing conditions, those patients demonstrated two conscious hemispheres operating in isolation from each other. To read the ensuing literature in the 1960s and '70s, one would never guess that the idea of the double brain was not a totally new discovery.

How are the "split-brain" of the 20th century and "double brain" of the 19th century different? Both initially relied on data from small clinical samples of individuals with brain abnormalities whether from injury or disease. However, modern laboratory experimental techniques not available to the Victorians have allowed for the study of hemispheric specialization in healthy participants. Research from both eras have been applied to various other fields, such as education and criminology, with the very same issues plaguing them. Can one "educate" or train a hemisphere, especially in a person with a healthy brain? Does the relative dominance of a hemisphere biologically dictate one's abilities and talents? Is there meaningful psychological significance, either for personality characteristics or intellectual ability, in being left- or right-handed or ambidextrous?

Over time, the right hemisphere increased in status, no longer the inferior half.[113] Bogen, the neurosurgeon for the original split-brain patients, had viewed the two hemispheres as complementary in function with the left hemisphere as propositional and the right as appositional, the latter processing information in a non-linear fashion.[114] In contemporary popular culture discourse, the right hemisphere, once thought to the feminine, irrational half, has now been connoted by some as the masculine, creative hemisphere. This may also reflect the widespread popular belief of females outperforming males in verbal tasks (and hence the association with the verbal left hemisphere) and males outperforming females in mathematical tasks.[115] There is still a tendency to view specialized hemispheric function in terms of dichotomies.

A few other parallels in the conception of the dual brain can be seen then and now. Gazzaniga has proposed an interpreter or executive function for the left hemisphere.[116] The left interprets experience for the largely non-verbal right, even to the point of concocting plausible narratives in the absence of direct perception. This view is similar to Hughling Jackson's notion of the left hemisphere having the higher order mentation and consciousness while the right had a secondary level of consciousness. One had a "knower and an object/subject" because every thought was activated twice, once by each hemisphere.[117] However, the individual only became conscious of changes in "selves" indirectly through the mediation of images and verbal symbols. Also, hemispheric compensation is now viewed as a given, much as it was in Broca's "principle of substitution."[118] Specialized functions can sometimes be taken over by the healthy hemisphere in response to damage of the other.

Finally, the advent of contemporary imaging techniques, such as the PET and fMRI scanning, has sparked intense interest in localizing brain function. These methodologies allow a glimpse into the differential action of the two hemispheres in ways only imagined in the past. However, they also lead researchers to fall into the same potential error of assuming that localization of symptom or behavioral/cognitive response equals localization of function. There has been an emphasis on finding discrete structures rather than considering networks and pathways. Much of our discourse on the split-brain is subject to the Homunculus problem. Just who is this "interpreter" and does the interpreter have its own interpreter, and so on? These problems are partly due to limitations in technology but also our philosophical renderings of how the brain operates and how we tackle the binding problem of consciousness.

More recently, likely because of technological developments that permit imaging of neural pathways, theories of consciousness have indeed focused on modules, networks, and emergent properties.[119] Pinto et al. examined data from split-brain patients for consistency with two of the competing modern theories of consciousness. Results indicated that despite the separate and independent perception of each hemisphere, responses were still integrated. "Local recurrent interactions between neural areas ... are enough to create consciousness, even if these interactions are not part of a larger integrated network."[120] Unsurprisingly, these notions are not unique to the 21st century either. Roger Sperry's concept of emergent interactionism was proposed 50 years ago.[121] Perhaps we will someday know definitively how our "double brain" works.

3. "To Sleep, Perchance to Dream"

SLEEP IS SEEMINGLY ALL THE RAGE—not that it's ever gone out of fashion, of course, being vital to our survival—but it's been rediscovered by neuroscience, big Pharma, and the health-conscious public. Good sleep hygiene is now a byword (or phrase) and an entire industry of portable home sleep aids has popped up to give us innumerable choices of devices to monitor our circadian rhythms. Even the field of sleep medicine has taken off, once the provenance of dubious and non-credentialed "therapists" in strip mall centers and now firmly entrenched in the medical profession establishment. With the advent of smaller, more powerful, and portable electronics, the recording paraphernalia of sleep medicine is even infiltrating the home.

But perhaps surprisingly when it comes to sleep hygiene, there is relatively little that is new when compared to the advice given by the Ancients: Galen, Hippocrates, Avicenna, to name a few. Certainly the theories of what causes sleep have evolved and we can now describe sleep from any number of physiological angles but the essentials of how to get a good night's sleep are largely unchanged over the centuries, barring the more modern strictures that deal with the advent of electricity into our everyday lives.

A third of our lives is spent in sleep and yet the phenomenon remains one of the steadfast mysteries for neuroscience. Perhaps no other neurological aspect of the nervous system has been so well described; yet the underlying mechanisms are still poorly understood. Once thought to be a "non" phenomenon, sleep did not even merit much in the way of medical or scientific discussion. It was simply a negative state: the absence of consciousness. Now we know that the idea of sleep as a passive state is completely erroneous and the chronic absence or disruption of any sleep has serious health consequences.

Contemporary neuroscience and introductory psychology textbooks all have a chapter devoted to sleep, sometimes by itself or sometimes in conjunction with didactic material on attention and consciousness. Sleep is viewed now as a particular state of consciousness with unique characteristics. What follows is a description of basic sleep principles as understood by modern neuroscience. Then I will take the reader on a journey to see how this understanding changed over time and how pathologies of sleep were both discovered and created.

For the normal, healthy adult, sleep involves of two basic phases with four to five distinct and recognizable stages, depending on how one counts them. The most basic distinction is between Quiet Sleep (or Non-REM) and Active Sleep (or REM, also known as "paradoxical sleep").[1] We begin our repose with Quiet Sleep, entering a light, drowsy Stage 1, characterized by theta waves. This stage progresses to Stage 2, identifiable by the presence of sleep spindles and a K-complexes, both being brain wave signatures obvious with an electroencephalogram recording. Stage 3 is the beginning of Slow Wave Sleep, sometimes called emergent or Early SWS. Here the previous theta waves begin to give way to increased amounts of larger, rolling waveforms called delta waves. By the time 50 percent or more of the waveforms are delta waves, the individual is now in the fourth stage, Late SWS. This is a stage of profoundly deep sleep from which it is difficult to rouse someone.[2]

The progression through the stages takes about 60–90 minutes, after which time a remarkable change occurs. The brain enters Active Sleep with the iconic appearance of REM, or rapid eye movements. Whereas previously still during the Quiet Sleep phase, the eyes begin to dart back and forth beneath the eyelids. The voluntary muscles lose their tone, becoming virtually paralyzed (atonia), and even more astonishing, the brain waveform returns to the pattern associated with being awake: beta waves![3]

This active period of the brain being "awake" while the individual is profoundly asleep gives rise to the moniker, "paradoxical sleep."[4] In 1967, Aserinsky and Kleitman famously made the discovery that dreaming mostly occurs during this stage.[5] Active sleep lasts for approximately 30 minutes and then the cycle repeats, typically four to six more times throughout the night. However, with each repetition, active sleep constitutes a greater proportion of the cycle, in fact, as much as 100 minutes. Thus, by the end of one's sleep with several cycle repetitions, an individual has a higher probability of awakening during REM than any other stage.[6]

Certain physiological and neurochemical elements of sleep and waking are thought to be understood today. Unsurprisingly, there are different mechanisms involved with quiet versus active sleep. Quiet sleep is guided by a hypothalamic sleep-wake "switch" that involves several areas within the hypothalamus in the basal forebrain area.[7] Antagonistic neurons in the VLPO (ventrolateral pre-optic nucleus) containing the inhibitory neurotransmitters, GABA and Galanin, operate reciprocally so that when these sleep-promoting neurons within the anterior hypothalamus are activated, wake-promoting neurons in the posterior and lateral hypothalamus and basal forebrain are inhibited or "disfacilitat[ed]."[8]

Ultimately, pathways projecting to the thalamus and cortex are inhibited so that there is in effect, an uncoupling between the brain stem and higher brain structures. This results both in decreased cortical arousal from ascending sensory tracts and cortical responsiveness from the descending motor tracts.[9] Oscillations in these tracts are reflected in the sleep spindles of Stage 2 and the synchronous delta waves in Stage 3. It has been suggested that these oscillations are responsible for the pruning of redundant neural synapses and memory consolidation.

Metabolic processes are diminished during Quiet Sleep, which supports the notion of non–REM sleep being responsible for the resting functions of sleep. Yet it would be incorrect to presume that the brain is actually "quiet" as PET scans reveal that 80 percent of the brain is still active during non–REM sleep.[10] The slow wave, synchronous activity reflects activity for some purposes. Quiet sleep is in fact more than just a decreased wakefulness that is suggestive of a continuum. Consciousness during sleep is altered "in parallel with the reorganization of brain activity."[11]

The neuropeptide hypocretin (or Orexin) mediates the transition between Quiet and Active sleep, as well as being active during wakefulness.[12] Normally, the release of hypocretin inhibits REM sleep; in individuals with narcolepsy, there is degeneration of hypocretin neurons, perhaps due to an autoimmune response, and thus REM sleep intrudes during wakefulness. The result is a "sleep attack," complete with cataplexy, a sudden loss of muscle tone.

Active or REM sleep also has "on" and "off" switches, located in the reticular formation of the brain stem (pons) and midbrain areas.[13] The pons has a REM sleep center near the Locus Coerulus. Cells here inhibit the motor neurons from the cortex to produce the muscle atonia, resulting in virtually paralyzed muscles. Glycine also inhibits motor signals from the spinal cord. Cortical neurons, previously firing in synchrony during the slow wave portion of Quiet Sleep, now show a desynchronized pattern akin to wakefulness. The REM-off neurons are found in the Locus Coerulus, which releases norepinephrine (chemically related to adrenaline). In 2005 Saper, Scammell and Lu proposed a "flip-flop" switch where a "circuit of mutually inhibitory elements sets up a self-reinforcing loop, where activity in one of the competing sides shuts down inhibitory inputs from the other side, and therefore disinhibits its own action."[14] The homeostatic balance is such that when either side begins to overwhelm the other, the switch "flips."

In contrast to Quiet sleep, metabolic processes are increased in the brain during REM.[15] Increased blood flow is directed to the visual cortex, while the prefrontal cortex has diminished activity. As the latter is associated with judgment and impulse control, the decreased blood flow perhaps is correlated with the uncensored nature of dreaming associated with REM sleep. The former, of course, is associated with the strong visual content of dreams. Hobson and McCarley's Activation-Synthesis model, originally published in 1975, proposes that dreams are a function of the activated cortex attempting to make sense of the neural activity in the occipital lobe and other cortical areas when the cortex has been switched "on" by the reticular activating system (RAS).[16] While the particulars have since been elaborated by the REM-on and REM-off switches, as suggested by Saper et al., the basic premise of the theory remains intact today.[17]

The above account of the physiology of sleep is undoubtedly simplistic; I have left out the laundry list of brain nuclei and neurotransmitters that interact in this incredibly complex phenomenon that constitutes sleep. The key points, however, should illustrate the current view that sleep is not a passive state but one that is active and one that is constituted by myriad neural networks. This view, however, is a relatively recent one and the knowledge gained about sleep processes has culminated in the subsequent development of a new medical sub-specialty, sleep medicine. In fact, sleep has had a complete make-over from its previous depictions in the scientific and medical communities.

Sleep was long considered simply a state of absence: that of not being awake or conscious. Indeed, the ancient Greeks viewed sleep akin to death as the god of sleep, Hypnos, was the younger brother of the god of death, Thanatos, both being the sons of Nyx, the god of night.[18] We still use the expression "dead to the world" in relation to someone in deep sleep. In fact, Plato admonished his students to sleep as little as possible.[19]

However, even the early Greek philosophers disagreed on a putative physiological theory of sleep. The earliest recorded theory was by Alcmaeon of Croton (fifth century B.C.E.), who speculated that sleep resulted from the brain filling with blood.[20] This "vascular congestion" theory dominated views about sleep for over two thousand years! Doubters included Hippocrates, who believed the opposite—that sleep came from blood leaving the brain. Aristotle espoused an atomistic theory related to digestive processes that was a type of congestion theory. Fumes from digested food were transported in the blood to the brain where they cooled. The cooled atoms then descended from the brain and when they reached the heart, the seat of sensory experience, the organ was enveloped by vapors and sleep ensued. Galen modified Aristotle's theory to focus on the brain and outlined the congestion theory that became the dominant vascular theory.[21] Sleep was one of the six "non-naturals," the Galenic categories that "operatively defined health or disease, depending on the circumstances of their use or abuse, to which human beings are unavoidably exposed in the course of daily life."[22] Passions of the mind and things excreted or retained are examples of other categories of the non-naturals. In all cases, they acted to balance the humors and thus were responsible for a relative state of health.[23]

Non-Western ancient views of sleep, such as those espoused in China, India, and the Middle East showed similarities in a holistic approach to the phenomenon. The circadian rhythms of sleep and waking indicated a mechanism to restore and maintain homeostasis for the organism, whether dealing with humors or some sort of organismic energy. For the ancient Chinese (and Traditional Chinese medicine today), sleep is seen within the context of yin-yang and the energy trajectory of Chi.[24] Avicenna, the renowned medical authority of ancient Islamic cultures, essentially had Galen's humoral approach and saw sleep as a necessity after excessive expiration during waking hours. The function of sleep was to restore balance to the four humours.[25] Practitioners of Ayurvedic medicine in ancient India outlined seven types of sleep, including "night" sleep (considered the healthiest) and "tamas" (ignorance; due to laziness and considered the unhealthiest).[26] In all these ancient traditions, good sleep hygiene was promoted and bears similarities to the dictums espoused today in contemporary sleep medicine.

As scientific theories incorporated new technologies through the ages, theories of sleep changed to accommodate those developments. For example, by the late Renaissance, the brain had replaced the heart as the origin of nerves and so theories shifted accordingly. For example, Descartes used the contemporary physics of hydraulics to propose that during sleep, the same mechanistic principles could be applied.[27] In *L'Homme et la*

Formation du Foetus, Descartes says sleep occurs when the "link between the brain and the outside world is interrupted by a contraction of the nerve tracts. Therefore, man can neither feel any sensations nor move his extremities."[28] When the nerves contracted, Descartes believed that the animal spirits in the pineal gland caused the ventricles to collapse. This mechanistic action induced sleep. The collapse was not total, however, allowing for the phenomenon of dreaming. The nerve fibers were relaxed in sleep but taut during waking hours.[29] During sleep the spirits would gradually be replenished and when the ventricles became full again, the individual awoke. Rest and restoration are the assumed functions of sleep here. Descartes famously wondered in his 1641 *Meditations on the First Philosophy* how we know that our reality is not in fact all a dream, dreams being a "creation of the great deceiver."[30] Consciousness is absent during sleep and so cannot confirm its own existence yet a dreaming mind is a thinking mind. However, during non-dreaming sleep, the mind cannot retain new sensory experiences and thus there is no recollection of events occurring during sleep, even with phenomena like sleepwalking.[31]

Because sleep was not viewed as a physiological phenomenon in its own right but rather a state of absence, little interest was seen in the study of sleep and so the basic vascular or congestion theory prevailed by consensus during the Renaissance and the Enlightenment periods. The notion of sleep as one of the "non-naturals" continued to be espoused by physicians but the mechanistic perspective of Descartes with its focus on the nervous system was beginning to take hold. Yet in a throw-back to Aristotle, sleep was still considered to be intimately related to digestion, triggered by the vapours moistening the brain. This view is evidences by writings of physicians, such as Cogan's 1612 *The Haven of Health* and George Cheyne in his 1724 *Essay of Health and Long Life.*[32] Tracts on sleep during these years tended to focus on sleep hygiene and the most common of sleep ailments: insomnia, nightmares, and somnambulism (sleep walking). Sleep was restorative "in which the efficient concoction of food into blood could take place, provided sleepers followed certain rules about moderating their diet, sleeping at night, not sleeping too much or too little, and minimizing their exposure to night-time vapours."[33] The London physician Thomas Willis had four chapters in his 1692 *The London Practices of Physick* on sleep disorders.[34] He agreed with Descartes' description of animal spirits but maintained that some of these spirits were active in the cerebellum, which was the basis for dreaming.[35] The spirits that escaped to the legs resulted in a condition we would today know as Restless Leg Syndrome.

David Hartley used his doctrine of vibrations whereby the mind could detect the vibrations of nerves to create sensation.[36] While still adhering to the congestion theory, he attempted to incorporate Newtonian physics in explaining how accumulated blood in the cerebral veins would exert pressure on the brain and thus induce sleep. Even if his particular idea of nerve vibrations did not catch on, it set the stage for examining sleep from a physiological perspective. In 1795 Johann Blumenbach observed a patient whose brain was partially exposed by an opening in the skull (trepanning) and noted that the brain was paler when the patient was asleep. He therefore concluded that sleep was due to a *decrease* of blood to the brain.[37]

This contradictory view would evoke puzzlement in those who espoused the mainstream congestion view. George Cabanis, in his 1802 work *On the Relations Between the Physical and Moral Aspects of Man*, also reported results of his empirical studies of the circulatory system during sleep and dreaming.[38] He noted that sleep was an active process, as evidenced by the increased blood flow to the brain. The nutriments of the blood wielded some influence on brain action, chiefly by an imperfect excitation especially during dreams. He likened dreams to madness.[39] Yet his studies did not include actual blood flow experiments but rather inspections of pathologies evidenced in the anatomy of brains.

Likewise, Robert MacNish in his popular 1842 *The Philosophy of Sleep* gives a thorough accounting of sleep using a phrenological framework for his congestion theory despite the noticeable waning of phrenology as a viable scientific theory.[40] He explains sleep and dreaming as differential states of particular cerebral organs, reflecting "complete" and "incomplete" sleep. "The former is characterized by a torpor of the various organs which compose the brain, and by that of the external senses and voluntary motion. Incomplete sleep or dreaming, is the active state of one or more of the cerebral organs while the remainder are in repose. The senses and the volition were either being suspended or in action according to the circumstances of the case. Complete sleep is a temporary metaphysical death, though not an organic one—the heart and lungs performing their offices with their accustomed regularity under the control of the involuntary muscles."[41] MacNish also observed trepanned patients and noted that during sleep the brain is sunken within the skull and paler. He also describes the outward manifestations of what we now call Quiet and Active sleep when he referenced the individual differences in "profoundness of sleep."[42] Individuals could differ by being light and deep sleepers but the distinction could also be seen within the individual during a single sleep episode. The first half of the night was generally spent in a deeper slumber—a result of the built-up fatigue—and the sleep became progressively lighter towards the morning hours as the brain is replenished. MacNish continues with observations of sleep that would be just as applicable today, as well as providing the descriptions of various sleep dyssomnias. Curiously, he even includes a chapter in his book on the sleep of plants!

The main challenge to the congestion theory came in the mid–19th century with the physiology experiments of William Hammond. Hammond was an American physician whose claim to fame was being surgeon general of the United States during the Civil War. He was also one of the earliest neurologists as that specialty was beginning to emerge as a medical subdiscipline. He espoused a theory of sleep that directly contradicted the congestion theory. Instead of vascular congestion, he proposed that sleep was caused by cerebral hyperaemia; in other words, decreased blood flow.[43] His supposition was based upon his empirical studies with trepanned animals where he was able to make systematic observations of the blood flow in the blood vessels in the dura mater covering the brain. He used a "cephalohemometer" to measure the ebb and flow of blood during sleep and waking.[44] Hammond expressed surprise that the vascular congestion theory had persisted in the face of contrary evidence but surmised it was due to equating sleep with being in a coma.[45] The latter was indeed induced by pressure of blood congestion in the brain and two states were similar in their consequence of loss of will or volition.

However, Hammond believed that the "immediate cause of sleep is a diminution of the quantity of blood circulating in the vessels of the brain, and that the *existing* cause of periodical and natural sleep is the necessity which exists that the loss of substance which the brain has undergone during its state of greatest activity, should be restored."[46] Reducing attention by reducing sensory stimulation meant that blood was diverted from the brain and thus resulted in sleep. The restorative necessity of sleep in synthesizing nutrients for the brain is apparent in this quote as well. He further notes that the brain cannot be active during the entirety of sleep because otherwise the brain would be exhausted. While the state of the mind during the sleep is predominantly one of absence, especially in regard to the sensorium, Hammond believed the entire brain was not always in repose. Some mental faculties were suspended and indeed some were activated through a loss of inhibition.[47] This "unconscious cerebration" explained not only dreaming, but the phenomenon of problem-solving during sleep whereby some mental activity was engaged without conscious awareness of doing so. And yet, problem-solving is rarely accurate while dreaming because judgment and will are inactivated. We may "imagine" that we do actions in dreams, but it is not true volition or will.[48] Interestingly, Hammond still adheres to the platonic tripartite division of the mind into Feeling/Emotion, Will, and Intellect and these are variously affected during sleep. Feeling is suspended, of course, due to the lack of sensation but emotion has free rein, being unshackled from Will. Some aspects of the intellect are intact during dreams, such as memory and imagination, but others like perception, judgment and reason are lacking.

Hammond's theory of cerebral anemia persisted, at least among physiologists, until the 1930s.[49] But in the 1880s, views of the medical and psychological community began to diverge from it. New sleep theories proliferated. Vascular, congestion, reflex, chemical, and neural theories were all vying for attention.[50] Neurologists and psychiatrists found a new focus on dreaming as a way to access the unconscious mind, while physiologists pursued sleep in the context of new research on muscle fatigue. Hammond had proposed that sleep was used to manage the build-up of waste products caused by mental or brain fatigue.[51] So, when Angelo Mosso suggested that the fatigue that led to sleep was analogous to the fatigue of exhausted muscles, physiologists were intrigued. The recent invention of the kymograph, the precursor to the modern day physiograph and polygraph, helped spur that line of research.[52]

At the same time, there was great interest in new developments in the medical world regarding hypnotism, which had some potential bearing on the dream-state.[53] The competing "schools" of hypnotism in Nancy and Paris, France led the research on using hypnosis as a diagnostic tool (Paris) versus as a therapeutic device (Nancy). The analogy of the hypnotic state with dreaming was an obvious one because volition was suspended during both conditions. Hippolyte Bernheim, co-founder of the Nancy school, ultimately disputed that similarity. Regardless, these attempts to explain dreaming were eclipsed by the 1899 publication of Freud's (1900) *The Interpretation of Dreams* (Freud had the publisher put the publication date of 1900 on the monograph despite the fact that it came out the previous year in order to be associated with the new century).[54] Within five years of its publication, with its focus on latent or symbolic meanings of dreams versus manifest content, dreams took on new meaning. Suddenly dreams had a purpose—not just memories of events—but as messages from the unconscious reflecting wish fulfillment.

But Freud's theory was essentially a biological one, positing the ebb and flow of instinctive energy transformed into psychic energy. While Aristotle had philosophized on the purpose of dreams as wish fulfillment, Freud was able to cloak the idea in scientific garb, incorporating evolutionary biology and then contemporary ideas of neurophysiology. Increased tension from repressed desires of the evolutionarily more primitive Id resulted in a build-up of pressure needing release. Dreams became the "safety valve" for the release of that tension.[55] Neural centers with subsequent nervous exhaustion then needed replenishment by the next bout of sleep.

Neural theories in the latter decades of the 19th century also took their cue from ideas about energy in the body. As notions of nervous exhaustion gained precedence in explaining the near epidemic of anxiety and nervous disorders, including hysteria and neurasthenia, sleep played an integral role in recovery. Nobel prize winner Ivan Pavlov, whose fame in the psychology and neurology fields was due to his paradigm of classical conditioning (the creation of associations between neutral stimuli and reflexive responses), applied his neural reflex theory to sleep.[56] Sleep was a function of inhibition of neural centers, a spreading of inhibition that gradually encompassed the cortex. "Excitation and inhibition, Pavlov argued, irradiated across the surface of the cerebral cortex, forming a 'mosaic' of points that guided the behavior of the animal at every turn."[57] Each center of excitation was surrounded by an area of inhibition and with repetition of the pairing of unconditioned and conditioned stimuli, the inhibition area was enlarged to the point of affecting the entire cortex. Thus sleep ensued as a means to protect the neural cells from the debilitating effects of fatigue.

In contrast, there were chemical theories, as exemplified by the Japanese physiologist Kuniomi Ishimori in 1909 and the French experimental psychologists René Legendre and Henri Piéron who proposed in 1913 that sleep was due to the presence of a hypnotoxin.[58] Coming out of the late Victorian fascination with pharmacology, toxicology, and the new developments in analytical chemistry, the idea that some chemical in the brain induced sleep was a reasonable one. However, Piéron was also part of the experimental psychology movement giving rise to behaviorism and comparative animal behavior in an effort to shift methodologies away from introspection. He built upon the hypothesis of Swiss physiologist Eduoard Claparéde that sleep must be guided by a "sleep center" in the brain. Piéron deprived dogs of sleep to the point of death (his method of "prolonged" or "enforced" wakefulness) and then extracted brain tissues and fluid, including cerebrospinal fluid.[59] Sleep deprivation was a "pathological mirror of sleep that could help illuminate the normal mechanisms of this mysterious state."[60] Those substances were then systematically injected into healthy dogs to examine the effects on sleep. When the cerebrospinal fluid was injected, sleep was induced. Piéron was able to show that the hypnotoxin was found only in the nervous system and proposed that the rhythmical build-up and exhaustion of the substance was manifest in the sleep-waking cycle.

The idea of a sleep center unexpectedly gained prominence when an outbreak of Encephalitis lethargica occurred between 1917 and 1929. In 1928 alone, there were 85,000 cases worldwide.[61] Constantin von Economo was a neurologist at the time who is credited with the "discovery" of the disease as a variant of encephalitis and named it Encephalitis lethargica. This is the same disease popularized in Oliver Sacks' book *Awakenings* and the movie of the same name. This sometimes fatal disease of unknown origin is characterized by excessive sleepiness—patients sleeping for weeks at a time. It is not the same as being in a coma in that the patients could be roused relatively easily; however, they could not stay awake. Von Economo's research indicated that a specific area in the midbrain (e.g., hypothalamus) was affected in the disease. This discovery became an important step in identifying the brain areas and mechanisms for sleep and waking.

Competing approaches to conceptualizing sleep continued to be promoted during the next few decades. Crichton-Miller (1930) outlined at least four different theories in his medical treatise on Insomnia: vascular (congestion theory once again), metabolic (citing Piéron's work on a hypnotoxin), neurological (positing localized sleep centers, perhaps in the hypothalamus), and reflexive (Pavlov's irradiation theory).[62] However, two independent strands of research in the 1930s began to converge. One line of research involved transection studies in which brain lesions were performed in discrete areas in order to examine effects of sleep and waking. The other major development was the discovery that the electrical activity of neural cells in mass could be recorded and displayed as an electroencephalogram—the EEG.

Brain wave recordings paved the way for the serendipitous discovery by Aserinsky and Kleitman in 1953 of Rapid Eye Movement (REM) sleep, which in turn led to the further differentiation of sleep stages and description of the sleep cycle.[63] According to Dement, who was a graduate student of Kleitman's and now one of the leading authorities on the topic of sleep, previous EEG studies of sleep had been performed using time sampling in order to save paper and allow for the researchers to take naps themselves during the night.[64] Since sleep was presumed to be a single state, there was no need for continual measurement. Kleitman's laboratory was focusing research on the physiology of the rhythmical nature of sleep.[65] While monitoring the EEG of a sleeping participant, Aserinsky (another graduate student in the laboratory) noted a change in the EEG that indicated the participant was suddenly awake.[66] When he went to observe the participant, to his surprise, the participant was in fact most definitely asleep. At first, Kleitman was not convinced that the participant had not awoken, thus resulting in the change in EEG pattern. However, they followed up by conducting a new study of 20 participants and now utilized EOG (electro-oculogram, or recording of eye muscles) and video evidence. A distinct phase of sleep was detected that was characterized by rolling eye movements and an "awake" EEG pattern. Of the 27 awakenings during this rapid eye movement stage (REM), all 20 participants reported having been dreaming. When participants were awakened from the non–REM sleep, they reported not dreaming in 19 of the 23 awakenings.[67] The importance of this landmark study, besides the description of REM sleep, was to show that dreaming occurred rhythmically.

While this study did not yield much fanfare at the time, Dement's follow-up studies proved to be ones that would shape the future of sleep medicine.[68] As a psychoanalytic psychiatrist, Dement's goal was initially to explore the physiology of dreams to support Freudian theory and he believed that the objective manifestation of the underlying physiology of REM sleep supported the notion of emotional release during dreaming.[69] Instead he pioneered the work that would reveal the architecture of sleep with its distinct stages.

Gottesman points out that the EEG studies of the 1950s and '60s (including Jouvet's work that demonstrated the concomitant lack of muscle tone during REM, which led to the term "paradoxical" sleep) paved the way for modern sleep research.[70] This assertion has merit in that the use of the EEG allowed one to peer inside the workings of the brain. "It was not the case that dreams protected sleep; it was rather the other way around, in that sleep's rigid architecture, invisible before the EEG, protected dreams."[71] Objective instrumentation has always held an allure to science. Sleep had been a phenomenon particularly inaccessible to research prior to the discovery of brain waves. The advent of "wet" physiology at about the same time allowed scientists to detect the biochemistry of sleep, especially that of the neurotransmitters such as serotonin.[72]

While no one theory had held traction during the early years of sleep research, each in its own way set the stage for the contemporary theories we see today. Perhaps the most important result was the shift from conceptualizing sleep as a passive state of absence to an active process. This change in thinking not only created a research climate for the study of sleep for its own sake rather than as an epiphenomenon, but led to the development of the field of sleep medicine. Now with the plethora of sleep disorders and specialized sleep clinics to treat them, it is rather remarkable that these were largely nonexistent a few decades ago. Sleep disorders that had been largely dismissed due to the reliance of patient self-report suddenly had legitimacy because there are now objective signs to be observed, thanks to the technologies to "see" into the brain.[73] However, with that shift has come a tendency for medicalizing sleep variants that had previously warranted little notice by health professionals. Williams argues that some of the sleep disorders now promulgated in pharmaceutical advertisements are socially constructed. What, in the public's mind, constitutes a "good night's sleep"? Subsequently, an entire "sleep industry" has emerged, complete with diagnostic codes, health care reimbursements, and a multitude of self-monitoring devices. A shift in medicine occurred from using "subjective symptoms to objective signs."[74] Once the purview of the family physician or even neurologist, sleep medicine has become the provenance of pulmonary specialists, thanks to the surge in sleep apnea cases. This discussion is not to minimize the potential negative health outcomes of untreated sleep apnea and severe insomnia; I only point out the privileging of certain kinds of knowledge, in this case technological (e.g., EEG versus self-report), to decide that one is having a poor night's sleep!

There is an ahistorical flavor to contemporary medical writings on sleep. It is as if various dyssomnias and parasomnias never existed until the last few decades even though many of these disorders were recognized in previous centuries. While recent years have brought forth new depictions of the physiology underlying these phenomena, there have yet to be definitive answers regarding the how and why of sleep. Our modern theories look suspiciously like the ones promoted in previous centuries but now our current labels are more scientific and authoritative.

Popular articles and books promoting good "sleep hygiene" abound in recent years. Yet these treatises, too, sound suspiciously familiar. The website for the Mayo Clinic offers the following suggestions: have a bedtime ritual, watch what you eat and drink beforehand, stick to a schedule for sleeping/waking, don't sleep with pets, sleep in a cool and comfortable environment, don't nap too long during the day, minimize stress, and exercise during the day. Among the behavioral treatments listed for sleep apnea, avoidance of sleeping on one's back is listed. These are all sensible suggestions derived from empirical research on sleep and insomnia. An inspection of historical documents on sleep hygiene reveals a familiar refrain, however.

Avicenna in his 1025 A.D. *Canon of Medicine* advises moderation in sleep with attention to timing and duration, moderation of what one eats before attempting to sleep, and discusses the interaction of insomnia with mental health issues.[75] Sleeping on one's back is discouraged, although not because of the physical obstruction of the air passage as is thought today but rather in order to distribute the humors evenly throughout the body. Ayurvedic medicine of ancient India has specifications for sleep hygiene, including sleeping on one's side rather than the back, guidelines for napping, and also notes the relationship of mental health and sleep.[76] Ancient Hebrews followed guidelines in the Talmud on duration of napping, avoiding alcohol before bedtime, and minimizing stress to treat insomnia).[77] The Middle Ages brought forth similar dictums about sleep: never sleep on one's back, naps must be short or preferably avoided, and one should not eat just before bedtime.[78] Cogan in his 1612 *The Haven of Health* also advised against back sleeping, napping during the day and cautioned his readers to "sleep in a cold place well covered."[79] The Enlightenment years brought similar advice from notable physicians, such as Thomas Willis, whose 1692 book on *The London Practices of Physick* contained four chapters on sleep disorders, including a description of what is now termed Restless Leg Syndrome.[80]

Medical texts and articles from subsequent centuries routinely contain sections on insomnia and various sleep disorders.[81] Sleep paralysis, the feeling of chest compression and source of the original incubus, "night mare" or "hag," was well documented beginning in medieval Europe. Gordon maintains that each generation has interpreted this REM dyssomnias according to the cultural "schemas" of their time.[82] Many treatises covered the topic of somnambulism, as sleepwalking not only had health connotations, it had implications for judicial matters as well.[83] Early pre–Baconian scientific periods viewed the phenomenon as demonic possession while in later eras, such as during the Enlightenment, it was viewed as a disease of the will. These accounts all contradict contemporary assertions that sleep disorders were disregarded by the medical community until the second half of the 20th century with the advent of modern sleep medicine.

Drug therapies for insomnia have also changed little throughout the centuries. Prior to the 19th century, opiates, particularly in the form of laudanum, were the most widely administered drugs for sleep. The Victorian medical bag had a broader arsenal of hypnotics, as sleeping agents were called. The bromides played a major role as sedatives but were joined in the mid–1800s by chloral hydrate, also known colloquially as a "mickey finn"—the knock-out drug. Chloral hydrate was the first psychoactive drug to be synthesized in a chemistry lab.[84] These addictive sedatives were widely used until the discovery of the barbiturates (e.g., Veronal [barbital] and Luminal [phenobarbital]) in the early 20th century.

Schulz and Salzarulo (2015) note that medical journal articles promoting pharmacological treatments for insomnia were about equal in number to those espousing behavioral therapies until the 1910s.[85] By the fourth decade of the 20th century, drugs had eclipsed suggestions for good sleep hygiene. This trend continued with the discovery of the benzodiazepines, especially triazolam (Halcion) in 1970. This hypnotic was thought to be much safer than the barbiturates, which were not only addictive but had a low threshold for a lethal overdose. Benzodiazepines are harder to overdose but have subsequently proved to be equally addictive. As with so many modern psychotropic drugs, there is a predictable cycle (the "Siege" cycle) whereby a new drug is hailed with great enthusiasm and optimism regarding therapeutic administration—in part, due to aggressive marketing by the pharmaceutical company, subsequently found to have negative side effects or unintended consequences, and finally followed by a more restrictive usage in the pharmacological repertoire.[86]

Zolpidem (Ambien), followed by eszoplicone (Lunesta), are two of the best known contemporary non-benzodiazepine GABA receptor$_a$-agonists that help initiate sleep.[87] Both were blockbuster drugs when they first emerged on the market, thanks to aggressive direct-to-consumer marketing campaigns by their manufacturers and zolpidem remains in the top 20 prescription drugs sold in the United States. Both were determined to be safer from the dangers of overdose and while longer term use may result in dependence, the addiction is not of the same severity compared to the barbiturates or benzodiazepines. However, zolpidem gained notoriety from the side effects of non-conscious behaviors while sleeping that some patients experienced and both have been associated with sleep inertia and increased likelihood of falls upon waking. Schwartz and Woloshin famously reported that the effectiveness of eszopiclone over a placebo was unimpressive, resulting in a 15-minute improvement in time to fall asleep and an increased duration of an average of 37 minutes.[88] To many with insomnia, however, this modest improvement and perhaps enhanced placebo effect is worthwhile.

So while much has been learned during the last hundred years to elucidate what goes on in the brain during sleep, the basic tenets are largely unchanged. Sleep remains a physiological imperative for the human condition with its periodicity and (usually) predictable cycles. The search for a "good night's sleep" has increasingly been an important medical, as well as public health priority and finding the contributors—as well as detractors—of that sleep continues to be a goal. An historical glance shows us that the behavioral descriptors of sleep and the behaviors that make up good sleep hygiene are mostly the same; what has changed are contemporary labels for the humors, the hypnotoxins, the vascular congestion, and the excitatory centers in the brain.

4. Bad Vibrations: Music on the Brain

IT WOULD BE HARD TO OVERSTATE the ubiquity of music in society. Indeed, one could almost consider it a universal of cultural expression and a fundamental characteristic of being human. Musicologists in the 19th century debated whether music emerged from language (another fundamental human characteristic) or naturally-occurring rhythms, ranging from one's own heartbeat (or mother's while in the womb), movements, and sounds in the natural environment.[1]

While music was studied by early experimental physiologists and psychologists in the late 1800s and early 1900s in the context of studying acoustics or sensation/perception, research interest waned during the middle decades of the 20th century. Hence, the neuroscience of music is a relatively recent phenomenon. Arguably the impetus for the resurgence of interest was the 1993 publication of a study purportedly demonstrating the so-called "The Mozart Effect."[2] Since then there are now thriving research programs in disciplines spanning topics in aesthetics to cognition to neurological disorders to therapy. There has been rapid, exponential growth in scientific publications on topics in the neuroscience of music during the last quarter century, leading some to view the neuroscience of music as a distinct sub-discipline in the field.[3]

Similarly, it is taken for granted that music has therapeutic or healing properties, as well as reflecting distinctive cognitive processing in addition to influencing brain processes in cognition and other domains. O'Kelly, Fachner and Tervaniemi note that 30 years of neuroscience research has revealed music's influence on virtually every aspect of human activity and that neuroscience is uniquely positioned to act as the academic area to unify such disparate areas of study.[4] Yet this was not always the case. In this chapter I will attempt to give an overview reflecting our basic understanding of the relationship of music and the brain and some of the many contemporary applications of that knowledge. Then I will take the reader to the "Age of Sensibility" when music had a far different reputation for its effects on the nerves. Remnants of those attitudes towards music were visible whenever "modern" musical genres have been introduced. Can we still see those vestiges today?

The relationship between music and the brain has been systematically studied in myriad ways, including trying to understand (1) where and how music is processed in the brain; (2) how listening and/or producing music might influence or alter brain function and consequently, behavior; and (3) applying that knowledge in therapeutic or educational settings. There is currently great interest in using music therapy to ameliorate symptoms of some of the most prevalent neurological diseases of aging, e.g., Parkinson's and Alzheimer's diseases.

As mentioned previously, studies on how music is processed by the brain harken back to the early days of experimental psychology and physiology. Hermann von Helmholtz (1821–1894) in the 19th century was a pioneer in studying the senses, particularly vision and hearing. Much of our foundational knowledge of acoustics and the physiology of auditory processing stem from his early work, published in 1863.[5] For Helmholtz, music perception was a function of processing fundamental tones and their compounds; it involved analyzing wave forms according to the amplitude (loudness), frequency (pitch), and unique quality (timbre).[6] As with any sound, Helmholtz proposed that nerve fibers in the inner ear (cochlea) resonated at the same frequency of the incoming sound wave. However, this type of acoustical analysis did little to acknowledge music's distinctive properties as compared with other sounds, including speech.

While Helmholtz is considered a pioneer of acoustical sensation and psychophysics, experimental psychologist Carl Stumpf (1848–1936) had a similar role in the study of musical perception. His 1911 work, *The Origins of Music*, is still today considered a classic in comparative musicology.[7] In it, he helped introduce the notion of music ethnography and placing music into cultural contexts. For Stumpf, it was not the individual tones that produced music but rather the arrangement of tones—in other words, it is the melody (which, by the way, can be transposed to different tones) that is primary in music.[8] Student researchers of Stumpf would go on to found the Gestalt movement in psychology, a school of thought that brought a more holistic approach to the examination of sensation and perception. While the perspective did not garner as much support among

American experimental psychologists as did the rival Behaviorist movement in the 1920s and '30s, the Gestalt principles of sensory perception are still a mainstay of every introductory psychology textbook.

Contemporary theories of the neural basis for music perception have been built upon this historic foundation. Using modern neuroscience research methodologies of fMRIs and evoked potentials, researchers have made great gains in our knowledge of how music is processed in the brain. Koelsch's 2013 neurocognitive model suggests seven distinct processes involved in perceiving music.[9] Auditory feature extraction is the first step where the tonal analysis, akin to Helmholtz's ideas, takes place. This is the "decoding of acoustic information" that occurs in various brain regions, particularly in the thalamus and auditory cortex, which have some tonotopic organization. This means that there are particular areas or locations in these structures corresponding to specific tonal frequencies. Other auditory features that are decoded at this preliminary stage include intensity or loudness.[10]

The next stage is Gestalt formation. Here the auditory cortex, particularly Heschl's gyrus and the planum temporale, processes musical features according to basic Gestalt principles: grouping and continuation.[11] We tend to perceive tones proximal in time as a group; we also tend to finish tonal sequences that are perceived to be a continuation of a musical line. Rhythm and melody are processed at a basic level here. Even the perceptual process of extracting figure from background (e.g., listening to an individual voice in a crowd) has its origin at this stage. This is then followed by an interval analysis that focuses on the relationship between tones and chords, as well as the time intervals. These two tasks are dealt with independently in the brain, as evidenced by case studies in which damage to a particular area can adversely affect the ability to perceive different tonal relations versus time relations in melodies. While the auditory cortex is involved in interval analysis, so is the basal ganglia and cerebellum—the two latter areas are brain structures associated with motor behavior.

Synaptic structure building is the fourth stage in Koelsch's model of music processing.[12] This is the neurological building of the musical syntax or grammar in the brain. Koelsch suggests that some of the same processes for language syntax are shared by its musical analog. The musical syntax involves the melody and phrasing, harmonies, instrumentation, and rhythms. According to Koelsch, this is a uniquely human cognitive process as well. The building of synaptic structures then goes through a phase of reanalysis and revision. What was first perceived to be the syntactical order may turn out to be a false lead. Koelsch gives the example of a tonal key perceived to be the "home" key, an assumption, in fact, that may turn out to be false.[13] Musical incongruities are also processed at this stage.

The sixth stage in this neurocognitive model focuses on the link between the perceived music and the peripheral nervous system, namely, the autonomic nervous system.[14] Vitalization occurs when the sympathetic and parasympathetic nervous systems react in the listener. The physiological response of musical "chills," changes in heart rate, blood pressure, respiration, and sweat gland activity, are all governed by the autonomic nervous system with its two divisions. Stress hormones, such as cortisol, are ultimately influenced by this system as well. Similarly, the immune system responds to boost levels of immunoglobulin A, resulting from its connections with the autonomic and endocrine responses. These physiological reactions are the mechanism by which music is thought to exert its therapeutic effects, ultimately lowering one's stress response, enhancing positive affect, and improving one's immunity.

Zatorre (2005) has also argued that music is uniquely suited for the field of neuroscience to explore numerous brain functions, ranging from acoustical processing to cognition.[15] And yet, the very complexity of music, as exemplified by Koelsch's cognitive processing model, gives hints to the enormous challenges inherent in such an endeavor.[16] Multiple areas of the brain are involved in music perception and likely both with parallel and hierarchical processing. Even with the technological advances in studying brain function, such as with functional imaging, the patterns of brain activity yield only small subsets of neural circuits. Parsing music into elemental stimuli and processes in a systematic fashion, however, has led to some insights regarding the "how" and "where" music is perceived by the brain. For example, Zatorre notes that while pitch is processed by both auditory cortices, the ability to discriminate among close pitches is better mediated in the right hemisphere.[17]

Just as interesting as the question of how the brain processes music perception is how music exposure changes the brain or "how the brain is sculpted by musical experience."[18] The plasticity of the neural wiring of the brain is evidenced by both anatomical and functional changes in the brain in response to intensive musical training, especially among children. Changes in auditory perception, motor areas, and neuron density have been identified in various studies. Zatorre also hypothesizes that music may affect individuals via mirror neurons as a mechanism for connecting music perception with motor responses, including emotion. Karmonik and colleagues note the potential value of music listening in improving brain functional connectivity between neural networks among patients of various neurological disorders, including stroke.[19] The results of their study indicated that the areas in the brain impacted by music are not just localized to music per se but rather underlie more functions that can serve more generalized cognitive processes. For example, attention, memory, emotion and motor responses are all influenced by listening to music; the same reward pathways (ventral tegmentum area and nucleus accumbens) that are activated by addictive drugs are also involved in the pleasurable response to music.

Margulis takes the relationship a step further in proposing that the experimentation and findings of neuroscience of music have shaped how music is viewed in its socio-cultural context.[20] The very fact that neuroscience currently has an authoritative label, deserved or not, has changed the way society now views music. There has been a subsequent "reification" of music's benefit on the brain and, by extension, health in general.[21]

Perhaps the most famous modern example was the sensation that surrounded the publication of "The Mozart Effect" in 1993.[22] Sporting a less catchy title of "Music and Spatial Task Performance," Rauscher et al. reported results demonstrating an improvement in 36 participants on some spatial tasks after listening to Mozart versus conditions of listening to instructions on relaxation or silence. The study was widely reported in the news and captured the general public's attention. It was not long before "The Mozart Effect" was fully entrenched in popular culture as an accepted "fact" with educational, political, and commercial institutions and enterprises capitalizing on this scientifically "proven" effect on learning.[23] Never mind the inconvenient fact that study by Rauscher et al. was seriously flawed and subsequent attempts to replicate the study had failed. The problems with the original study have already been outlined by others but demonstrating the finer points of research design, such as the lack of counterbalancing of conditions and conflating distinct cognitive tasks, has done little to stem the flood of Mozart Effect applications.[24]

Perhaps more importantly have been the subsequent research studies to tease out what brain processes may in fact underlie the benefits of music on cognitive performance. Schellenberg, along with colleagues involved in several previous studies, has provided painstaking evidence that enhanced arousal and positive mood changes elicited by music are likely responsible for any improvement in cognitive performance—not the genre of the music itself![25] With the mood effect, there is an important interaction with music preference. As Blood and Zatorre have shown, the idiosyncratic response of musical "chills" is based largely upon musical preference and is accompanied by changes in the addiction neural pathways and limbic system.[26]

So, while the Mozart effect may not have been quite what was initially advertised, there is still largely consensus among contemporary neuroscientists that music potentially has beneficial properties that can be applied therapeutically to various neurological and psychological conditions.

But was this belief always the case throughout the ages? Indeed, the benefits of music have been extolled by physicians and philosophers alike since the times of Plato and Pythagorus, and most likely earlier. The beneficial effects were originally related to the harmonic resonance and mathematical order seen in the cosmos.[27] The microcosm of the body was viewed by the ancient Greeks as a reflection of the macrocosm of the society and ultimately, the universe. Music that adhered to certain orderly relationships could align body and soul in a healthy manner. This view in health and medicine prevailed for centuries. "Each generation since the Renaissance, it seems, produced new reasons for music's healing powers."[28] The more interesting question to ponder is, however, has music ever been considered pathological? Not only has the answer been "yes" by the medical authorities, the underlying theme to that pathology also recurs with some regularity throughout history.

What changed in medical thinking about the malicious effects of music was a new model of the nervous system that arose during the Enlightenment. According to Kennaway, there had already been a shift with Cartesian philosophy from a focus on cosmology to the body itself.[29] Certainly, the ancient Greeks had discussed the problem of disharmonious combinations in music, especially modern music, but this represented a moral problem rather than a physiological one. A new emphasis on the role of nerves had emerged during the 18th century and in the latter half of the century, medical authorities had "identified nerves as responsible for music's emotional impact."[30] The musical vibrations translated to vibrations in the nerves. Still, music was a "model for order, morality and health ... a means of refining the nerves and of calming unhealthy passions, including sexual ones."[31]

During the 1700s, nervous diseases and the physicians who treated them were becoming medicalized and professionalized, respectively. A key work outlining the negative effects of stimulating the nerves was published in 1733 by the English physician George Cheyne.[32] *The English Malady* was a critique of modern life, and in it Cheyne argued that many illnesses were the product of the stress of modern life on nerves. This theme would be echoed again a century later. The right kind of music, Cheyne thought, could bring order to nerves and complement their action.

Conversely, stimulation of the nerves by music could potentially bring disorder to the body. An eminent Scottish physician and contemporary of Cheyne, Robert Whytt in his 1768 essay *Observations on the nature, cause, and cure of those disorders which are commonly called nervous, hypochondriac or hysteric*, also believed that music was mostly beneficial to health but if it overexcited the "passions," pathological effects could occur.[33] These influential works, along with Richard Browne's (1729) *Medicina Musica* helped establish the link between music and the nerves.[34] "Music's physicality, which had previously been the subject of moral concern, was re-categorized as a medical issue."[35] This view persisted throughout the 18th century and yielded a new concern about the pathology of music.

Browne had posited that music created a sympathetic vibration in the auditory nerve.[36] Since nerves were analogous to strings on an instrument, the vibration was a direct physiological response to the music. Refined and gentile individuals had a sensibility such that their nerves responded readily to the effects of music. Overstimulation would result in excessive sensibility and thus create a pathological effect that was deleterious to the nerves. Certain music, especially modern music, could have this overstimulating effect much the same way coffee or tobacco could.[37] It was not long before music was viewed as having the potential to harm the individual, especially females who suffered from excessive sensibility, rather than refining the nerves. Working class individuals were less vulnerable to the ill effects of music than upper-class individuals because their nerves were simply less refined in the first place.[38]

The stimulation of nerves soon shifted from a vibration theory to an electrical theory, using the new discoveries of galvanic electricity. Instead of literal vibrations, music was thought to impact the electrical fluid of the nerves. These views were further reinforced within a broader medicosocial context at the end of the 18th century with Brunonianism, a theory that all disease was a function of either under- or over-excitability of the nervous system.[39] There was "only one illness and that all health and sickness could be measured on a 'barometer.' Each individual had a limited stock of 'excitability' ... creating either 'sthenic' or 'asthenic' illness."[40] Thus, the common nervous disorder that took such precedence during the 19th century, neurasthenia, was essentially a weakness of the nerves. Kennaway points out that even with the fall of Brunonianism from medical discourse, the basic notions of over- or under-excitability of the nervous system persisted in explaining the contrary effects of music on health and illness.[41] This accounts for why there were individual differences in responding to the same piece of music: an individual would naturally have a distinct temperament that was prone to either increased or decreased excitability of the nerves.

Females were thought to be particularly vulnerable to the pathological effects of music due to their inherently weaker nervous systems.[42] In the late 1700s and early 1800s, as the number of pianos in the home proliferated, so did health warnings to women about the dangers of over-indulgence. At the same time, music and latent (or not so latent) female sexuality were conflated—there was "serious, masculine music relating to the transcendental subject and supposedly feminine, sensual music that merely stimulated the nerves."[43] Females were both victims of pathological music, suffering from neural exhaustion, fatigue, and poor reproductive health, and unwittingly the cause of pathological music in males when female sexuality infused the music, leading to a loss of nervous energy, self-control, and degeneracy.[44] Music was even proposed to be the cause of earlier menarche experienced in girls of greater socio-economic status—as opposed to greater nutrition and health conditions—because of its connection with the reproductive system. The respected German physiologist Johannes Müller believed that music affected nerves directly precisely because it was sound without the formal structure of words.[45] Music had the ability to excite the passions. Hermann von Helmholtz, Müller's student, expanded on this notion. Because the vibrations of music directly impacted the ear and needed no extra cognitive process to affect the passions or emotions, it uniquely had the ability to affect the nerves. The peculiar anatomy and physiology of the ear, as understood in the 19th century, allowed for an interaction of music and nervous system.[46]

The development of neuro-acoustics, as espoused by Helmholtz, happened in parallel to the burgeoning field of aesthetics. Eduard Hanslick, a contemporary of Helmholtz, was considered one of the most prominent musicologists in the Western world.[47] Hanslick argued that aesthetically pleasing music was differentiated from pathological music by its absence of feeling or affect.[48] The evocation of emotion from music was "antithetical" to the contemplative, rational response to music.[49] Interestingly for this discussion, Hanslick apparently realized that the current theory of overstimulation of the nerves by pathological music as the root cause was an inadequate explanation. In the 1850s, there was little scientific evidence for the understanding of a physiological response of music, sound, or even a biological basis for experiencing emotion, let alone how those phenomena were linked in the nervous system. Without an understanding of why particular harmonies or rhythms were disturbing to the nervous system, the aesthetic view was merely a philosophical one. Helmholtz's work on the physiology of audition and psychophysics was important precisely because it focused on the intersection of physical sensation and psychological perception. For Helmholtz, the notions of consonance and its counterpart, dissonance, were decipherable through the anatomy and physiology of the ear, creating a proto-neuro-aesthetic field.

The glass harmonica, invented by Benjamin Franklin in the 1760s, was the instrument that best exemplified this curious relationship of music and gender. Hadlock (2000) calls it the "perfect instrument for the 'age of sensibility'" due to "the immediacy of its effect on the listener—its ability to produce a spontaneous sensuous response."[50] The unique timbre of the sound produced by the glass harmonica was particularly associated with feminism and it was one of the instruments of that era that ladies could perform in public.

However, the instrument was also associated with madness due to its eerie tone and the direct vibrations through the fingers that would disturb the nervous system. The different musical tones were achieved by friction of the player's fingers on the rims of differing sized glass bowls that rotated. Franklin's invention was a variation of the notion of rubbing the rims of glasses filled with different volumes of water. This improvement allowed the performer to be in physical contact with more pitches, thus being able to play more complex musical compositions. It is no coincidence that Donizetti wrote the solo accompaniment in the opera *Lucia di Lammermoor*'s mad scene for the glass harmonica, though it is usually played today by the flute. Again, the theory behind the pathological relationship dealt with overstimulation of the nerves. According to Karl Röller in his *Über die Harmonika, ein Fragment*, "Overindulgence, creating excessive sensation with no opportunity to discharge it, will upset the body's equilibrium and damage the system."[51] The direct contact of the fingers on glass meant the vibrations were transmitted through one's body. Someone who practiced and performed to excess was in danger of permanent damage to the nervous system. "The glasses' persistent vibrations might overstimulate the player's hands, ears, and nerves, and might eventually overheat the warm blood that circulates in the fingers to produce a good tone. Prolonged exposure to the penetrating tone that initially lifted the spirits would induce melancholy and irritability."[52] Fingers were thought to be strong conductors of the "negative" electricity and the harmonica's sound could cause "prolonged shaking of the nerves, tremors in the muscles, fainting, cramps, swelling, paralysis of the limbs and seeing ghosts."[53] No wonder the instrument was actually banned in some European countries!

This link was purported to be the cause of death of one of the most famous performers of the day, Marianne Kirchgasser, and depression or melancholy in another, Marianne Davies. Ultimately, the glass harmonica fell from favor, becoming associated with mesmerists and charlatans.[54] Anton Mesmer used the instrument to set the ambient mood for his hypnosis demonstrations; but while magnetism as advocated by Mesmer, manipulated the electrical current of the nervous system, the sounds produced by the glass harmonica affected the nerves directly.[55]

Thus, during this era music could potentially be pathological to the listener due to the influence of sound vibrations on the ear and from there be transmitted to the nervous system via the auditory nerves emitting from the ears; alternatively, music could be pathological to the performer through direct transmission of sound waves to the nerves through the body—hence the term "Le corps sonore."[56] Musicians, in fact, were thought to be at greater risk for nervous diseases, such as melancholy, hypochondria, hysteria, neurasthenia—even infertility—all due to the malignant effects due to the physical effort of making music. The Romantic era in Western music, as exemplified by Beethoven, Brahms, the Schumanns, Schubert, and Mendelsohn, brought with it a tension of music to incite the passions and possible deleterious effects. No longer was serious music only about order but rather its ability to elicit feelings.[57] This relationship is apparent in E. T. A. Hoffmann's story *Rat Krespel* (*Councillor Krespel*), which became one of the stories featured in Jacques Offenbach's opera *Tales of Hoffmann*. The overexcitement of the passions induced by singing ultimately kills Antonia, Krespel's daughter. To modern day listeners, this scenario is typically viewed as implausible and gothic fantasy (particularly because in the opera, the ghost of the girl's mother incites her to sing) but when seen from the views on music and the nerves during the author Hoffmann's era, the plot is less fanciful and would have appeared to be altogether credible to contemporary audiences.

Fears about pathological music at the turn on the 19th century eventually faded, only to resurface periodically whenever new forms of music emerged in the public scene. The music of Richard Wagner in the mid–19th century is a case in point. His music was considered "both a symptom and a cause of nervous modernity and its physical and sexual pathologies—fatigue, 'neurasthenia,' perversion and degeneration."[58] The notion of inherited tendencies towards degeneration gained popularity among the medical community during this period and Wagner's music was thought to be a stimulus that could elicit latent tendencies just as other elements of modern life, including overuse of drugs, could. Wagner's music, above all others, was thought to produce an abnormal amount of nervous strain and in those with constitutionally weak nerves, neurasthenia or nervous exhaustion would result. In the most severe cases, homosexual behavior might emerge due to the perceived eroticism of Wagner's music.[59] The famous Victorian sexologist Havelock Ellis thought music and homosexuality were related. Sexual excitation without appropriate consummation was a leading cause of sexual neurasthenia.

Ironically, there would be a complete reversal of attitudes 40 years later when Wagner's music was co-opted by the Nazis in Hitler's Germany as epitomizing masculinity. Now not considered to be "new" music, Wagner's compositions were no longer viewed as dangerous to one's health. Instead the focus now turned to the new musical genre of jazz which held connotations of racial overtones and degeneracy. What Wagner's music and jazz had in common, however, was that when they emerged upon their respective musical scenes, they had wreaked havoc with conventional rules of harmony and rhythms. The discord was not simply due to disagreements within contemporary aesthetics and notions of "beauty" in music; those same controversies also played into the respective contemporary views of the inner workings of the nervous system. Wagner's pieces had "weak" rhythms, for example. His music constituted "nervous music" due to its draining of nervous energy, leading to various types of neurasthenia. The music could result in fatigue and weaken cognitive functioning, especially among those unfortunate individuals with degeneracy in their family lineage.[60] Furthermore, Wagner's music represented the downside and dangers of modern civilization. Just as the industrial age contribute to the age of anxiety because of the increased pace of society living, so did modern music with its penetrating effects on the nerves.

In contrast, the rhythms espoused in jazz overstimulated the nerves. The distinctive features of jazz music, syncopation and polyrhythms, eschewed conventional rules of music and represented a clear threat to morality and health.[61] According to Johnson, it was believed in the 1920s that jazz music not only excited the nervous system, it could erode mental functioning and shorten one's life.[62] Some critics believed the music worked to decrease mental intelligence or produced insanity by destroying brain cells. The evidence cited was that supposedly 50 percent of asylum patients were "jazz fiends." Much as the case seen a half a century earlier with Wagner's music, jazz represented the "unrhythmical noise" of modern civilization. In contrast, healthy music had regular rhythms: "right music rightly presented."[63] That was a tenet of the early professional societies of music therapists following the first world war. While eurhythmic music, by definition, was pleasing and had orderly rhythms to entrain the bodily functions, jazz music was chaotic and violated basic norms of music convention. "Surprising numbers of people, many highly educated, believed that an epidemic of defective jazz music and dance was threatening to produce a restless, 'nervously debilitated,' physically ruined, and mentally impaired nation."[64]

The idea of pathological music persisted and with each new musical innovation, a specter of potential doom for civilized society has been raised. The dawn of rock music brought forth concerns about the pathological effects on teenagers. Just as reactions to jazz music were tinged with latent sexuality, political and social upheaval, so were the same fears evoked by rock music. The strong connection of rock music and the social and sexual revolutions of the 1950s and '60s resonated with claims of the music adversely influencing the developing brain. As new technologies to deliver music were developed—phonograph records, compact discs, and even music videos, the discourse on pathological music has not been so different from that seen in the late 1700s. Kennaway notes that the earlier time in Europe when the notion of pathological music was prevalent was also a time of social and political revolutions.[65] Similarly, jazz and rock music both came into popularity during periods of social and political unrest. The main difference has been the changing views of nervous system function to rationalize why the new music is pathological.

The relationship of music and nervous system has mostly been benign throughout the last few centuries but punctuated with periodic outbursts of concern. In modern neuroscience, the only manifestation of purely pathological music is that of musicogenic epilepsy, a relatively rare phenomenon in which music is found to evoke epileptic seizures.[66] First described in 1884 and subsequently followed by the first clinical case study notes in 1937, this sub-type of epilepsy occurs in a small number of patients where complex music triggers focal seizures. Curiously, the patients in these documented case studies have all been described as "musically talented."[67] Why complex music, rather than simple tones, develops as a trigger is largely unknown but the suspicion is that it is the emotional response to the music, rather than qualities of the music itself, that acts more specifically as the stimulus. The temporal lobe and pathways through the limbic system of the brain are involved, as evidenced by abnormal EEG and imaging patterns.[68] These structures are all involved in the experience of emotion and emotional memories.

History of the relationship of music and brain shows us that new forms of music, as well as perhaps newer technologies to deliver it, will engender fears about influence on the brain and behavior. However, the history also shows us that those fears are often symptomatic of underlying assumptions of gender, race, and class during periods of social upheaval. It is now more likely, perhaps, that the new insights on the neuroscience of music that emerge with some regularity will be used to explain the beneficial aspects of music. The ability of music to assist the palliative or healing processes, whether for psychological or neurological disorders, has been known for millennia and contemporary neuroscience can begin to elucidate the mechanisms behind those processes. As a relatively inexpensive and non-invasive therapy, music has more potential for co-optation to benefit the nervous system rather than exhausting it.

Drugs and Behavior

5. Victorian Explorations of Psychoactive Drugs

Several years ago, when I was developing a new course on Drugs & Behavior, I explored general sites available on the Internet about psychoactive drugs in order to get a sense of how these drugs were being portrayed to the general public. Many sites were not academic in the slightest: www.veryimportantpotheads.com is a typical example. This particular webpage is one that could be categorized as sites that glorify or romanticize recreational psychoactive drugs. What caught my attention was the frequent referencing of celebrities, not just those of modern times, but historical celebrities who had used these drugs. They were being used as contemporary tools for propaganda of a particular social agenda. The celebrities from the history of psychology who were most often mentioned are Sigmund Freud, William James, and Havelock Ellis.

This chapter will take a closer look at these individuals and their use or experimentation with psychoactive drugs. Freud, James, and Ellis all had well-documented episodes with psychoactive drugs. Instead, I would like to examine briefly each case to put their drug use in a larger context with an aim towards determining whether the current exploitation of their celebrity status is justified by the evidence. Furthermore, their own explorations with these drugs illustrate the 19th-century views of nervous system action in general and psychopharmacology in general.

Sigmund Freud (1856–1939) had both a scientific and personal interest in cocaine that lasted many years—in fact, far longer than many holding the Freudian scepter initially admitted. As a young research scientist in 1884, Freud was engaged in microscopic neuroanatomy research, examining the nervous system of eels, and had very little clinical experience under his belt. He had apparently been reading articles about cocaine.[1] Of particular interest was an 1883 account by Theodore Aschenbrandt, a German army surgeon who wrote about six case studies regarding the miraculous rejuvenation of exhausted Bavarian artillerymen given cocaine in the field.[2] Freud also read several articles in the *Therapeutic Gazette*, a publication vehicle of Parke-Davis pharmaceutical company. These articles were little more than paid advertisements or testimonials, but when Freud read an 1880 article by Bentley suggesting that cocaine might be used to treat opium or alcohol addiction, he wrote the company to request a sample.[3] In fact, it was typical for George Davis of Parke-Davis to particularly solicit articles from physicians to show the efficacy of using their "tincture of coca" to cure morphine addiction.[4] This practice was commonplace in the relationship between the early pharmaceutical manufacturers and physicians in the field. Since the discovery of cocaine by Albert Niemann in 1860, there had been virtually no scientific research on the drug until Freud took an interest.[5]

At Vienna's Physiological Institute, the place where Freud had trained and was currently working on research projects part-time, Freud had a friend and colleague who suffered from a debilitating morphine addiction. In a letter to his fiancé, Martha Bernays, dated April 21, 1884, Freud writes about his intentions of experimenting on cocaine and alludes to Ernst von Fleischl-Marxow's situation.[6] He also remarks that perhaps his research will come to naught, but in the spirit of "nothing ventured, nothing gained," he is going to order some of the substance. At this time in his career, Freud had less than a year of clinical experience and no experience at all in treating addiction but it was not unreasonable to think from a theoretical standpoint that one could counteract the actions of a central nervous system depressant with a stimulant.[7] Even today the lay person often assumes that depressants and stimulants taken together will cancel out the effects of each. Instead, the effect of the combination is more often biphasic.[8]

So, in May 1884, Freud administered cocaine on a daily basis to his friend Ernst von Fleischl-Marxow as well as trying it himself. Merck, the European supply company, assumed that Freud and von Fleischl-Marxow were collaborating on research and published several company-sponsored pieces concerning the work.[9] Freud himself summarized the research in his 1884 monograph *Über Coca* and not only mentions Merck as the supplier, but also how expensive the cocaine was.[10] This caused the pharmaceutical company Parke-Davis, the American supplier of cocaine, to offer Freud reimbursement (60 guilders or ~$24) in exchange for his endorsement in 1885. Due to issues related to shipping, Parke-Davis was entering the cocaine market as Merck was ending its involvement. As an American-based firm, Parke-Davis had the ability to process the coca leaves more quickly before they lost their potency after being shipped from South America. Freud agreed to the endorsement and so entered into a financial relationship with Parke-Davis. According to Karch, Parke-Davis routinely quoted Freud's research as "proof that cocaine was an effective treatment for morphine addiction. No one ever mentioned that Freud first got the idea from reading Parke-Davis & Company promotional materials."[11] Nonetheless, after Freud's publication, both drug companies experienced a dramatic increase in sales of cocaine.

For Freud himself, cocaine proved to be the antidote for the episodes of depression that plagued him and he enthusiastically promoted the drug to his colleagues, friends, and family. He periodically sent samples to his fiancé with instructions for its use.[12] There is the now famous passage in his letter of June 2, 1884, to Martha Bernays extolling the effects of cocaine: "Woe to you, my Princess, when I come. I will kiss you quite red and feed you 'til you are plump. And if you are forward, you shall see who is the stronger, a gentle little girl who doesn't eat enough or a big wild man who has cocaine in his body. In my last severe depression I took Coca again and a small dose lifted me to the heights in a wonderful fashion. I am just now collecting the literature for a song of praise to this magical substance."[13]

Freud also conducted cocaine experiments on himself, which were included in the monograph *Über Coca* and published a few years later in journal articles.[14] In his monograph, Freud recommended cocaine as a panacea. It was not only a stimulant useful for lifting mood and specifically, depression, it could be used as an aphrodisiac, a treatment for nerve exhaustion or neurasthenia and cachexia (weakness), various gastrointestinal disorders, and as a drug enhancement for endurance. The latter use might be particularly useful for mountaineers or soldiers. A novel and unique effect of the drug was also to increase tolerance to the toxic substance, mercury. This was important because mercury was the only widely accepted treatment in the 19th century for syphilis.[15]

Freud's subsequent laboratory experiments focused primarily on the effects of cocaine on muscular strength (perhaps giving rise to his commentary on relative strength in the above passage to his fiancé) and reaction time. Occasionally, he recruited colleagues as participants. Karl Koller, a colleague at the Physiological Institute, was one of his assistants and later went on to conduct his own experiments with cocaine to examine its potential as an anesthetic.[16] It was his research that led to the discovery of using cocaine as an anesthetic for the cornea of the eye—a breakthrough for eye surgeries and a discovery that brought him international fame, much to Freud's chagrin. Freud had noted the anesthetic properties of cocaine but had failed to see the significant application of that finding to the eye. He later blamed the oversight on his fiancé because he went out of town to visit her instead of working in the laboratory when the discovery was made.

The sad outcome of Ernst von Fleischl-Marxow's cocaine and morphine use has been reported by many scholars.[17] Briefly, far from curing him of his morphine addiction, von Fleischl-Marxow was now addicted to both drugs, suffering from paranoid delusions and tactile hallucinations, known as formication, typical of stimulant overuse. These hallucinations typically take the form of believing there are insects crawling just underneath the skin, leading to constant picking and scratching. He was taking huge daily doses of cocaine at a time when Freud was publicly stating that cocaine was not habit-forming and that von Fleischl-Marxow no longer needed the drug. In fact, it was von Fleischl-Marxow's penchant for requesting very large orders of cocaine that led to the misunderstanding by Parke-Davis on the matter of collaborative research with Freud.[18] Von Fleischl-Marxow apparently had led the director of the company to believe that such a collaboration was in existence. In June 1885 Freud and others helped von Fleischl-Marxow through a life-threatening drug overdose crisis even as his second article on cocaine was in press.[19] Freud was concerned enough at his friend's addiction to cocaine that he warned Martha in a letter dated June 8, 1885, to be careful in using the drug.[20] Von Fleischl-Marxow died six years later—still an addict.

After an initially cool reception of *Über Coca*, physicians began to take up Freud's recommendations on treating addiction and publishing accounts of their own case studies, the results of which were decidedly mixed, if not downright negative. However, others of his medical colleagues were outright hostile in their response. Distinguished psychiatrist Albrecht Erlenmayer published a scathing indictment against Freud's treatment plan: "This therapeutic process has lately been publicly trumpeted and praised as a veritable salvation. But the greater the fuss made about this 'absolutely precious' and 'totally indispensable' route to health, the less efficacious it proved to be ... it was simply a question of propaganda expounded by individuals without any truly scientific experience, as objective analysis of the question easily demonstrated."[21] Erlenmeyer two years later went on later to blame Freud for releasing the "third scourge of humanity," after morphine and alcohol![22] By 1887 the dangers of cocaine use and addiction were well documented in the medical literature and the leading physicians of the day were counseling against the use of cocaine to treat opiate addictions.[23]

Freud's (1887/1963) response was a defensive article, blaming others for incorrectly administering the drug.[24] In that assertion, he conveniently ignored that critics were following his own 1884 instructions outlined in *Über Coca*. Furthermore, he suggested that cocaine was addictive only in persons already dependent upon other substances. That statement is also nonsensical given his earlier assertions that one of the main therapeutic purposes of cocaine was to treat morphine addiction. This 1887 article was his last public statement about cocaine in the medical literature, although he continued his personal use for at many more years, at least through 1895, to treat his depressed mood states, migraine headaches, and nasal problems.[25] While it can be questioned whether Freud was himself ever truly addicted to the drug, he does report symptoms in his correspondence at times that are quite consistent with chronic use, such as heart arrhythmias, insomnia, and paranoia. Interestingly, Freud himself in a 1908 letter to Carl Jung mentions paranoia as an outcome of chronic cocaine administration.[26]

So here we have a case of a young, ambitious, inexperienced clinician who rather hastily publishes suspect and sloppy research results (note he omits information on the participants, dosages administered, treatment duration, and even the dependent variables that he uses to measure effects), stakes his career on a disastrous premise of using cocaine to treat morphine addiction, and is accused of releasing the "third scourge of humanity" in the world! At least two researchers speculate that Freud could not have even read the actual articles from his own reference citations on cocaine because of the number of errors in his literature review and misunderstanding of the extant literature.[27] Is it any wonder he abandoned this avenue of research under the crush of negative critique and went on to study hypnosis in Paris with Charcot? As a postscript to this episode, Freud wrote to Martha Bernays in January 1886 during his Parisian trip that he used cocaine to boost his confidence in the presence of the eminent physician, Charcot.[28] His next big publication—and the one that would actually lead to his international fame was *The Interpretation of Dreams*.[29]

William James (1842–1910) represents an altogether different case of drug use. William James, brother of famed "gilded age" author Henry James, began his studies in medicine but abandoned the thought of applied medical practice and went on to become one of the founding fathers of modern psychology, ending his career as a professor of philosophy. Perhaps one of the great geniuses of his time, his contributions to psychology and philosophy are too numerous to list for this chapter. His interests were manifold and spanned questions of perception (perhaps connected with his passion for art), consciousness, and ultimately religious and spiritual experience. He is known in psychology for establishing the first experimental laboratory in the United States (that is, if one counts a closet with equipment!) and the first best-selling textbook of psychology: *Principles of Psychology.* This text is still popular today in the field and James' insights into various psychological concepts is amazingly prescient, indicating a keen observer of human behavior.[30] However, James eventually abandoned psychology, disliking the increasing emphasis on experimentation, and moved full time into philosophy. Jacques Barzun writes in 1983 that James had a view of the unconscious such that "the line of demarcation from the conscious is dim and shifting; that the unconscious is at once individual, collective, and possibly wider still."[31] Throughout his life, this artist-philosopher-psychologist was interested in "exceptional mental states."[32] He was intensely interested in the spiritualist movement that was overtaking the country and as a member of the Society for Psychical Research, helped lead investigations on one of the most famous seers, Mrs. Piper. Croce states that James "wanted to pursue his curiosity about the human mind without reference to authorities, assumptions, or prior faiths."[33] James' most famous work in philosophy in 1902, *On the Varieties of Religious Experience,* is a testament to his curiosity and keen interest in examining the limits of human consciousness.[34]

Surprisingly perhaps, the emerging field of neurology was considered a legitimate venue for experimentation on supernatural phenomena.[35] Neurology could establish a scientific, empirical approach to topics that had been the provenance of religion or superstition. During the latter part of the 19th century, neurologists used "a materialist assumption to claim special understanding of a wide variety of human problems."[36] These problems included trances, hypnotism, somnambulism, automatic writing, and dual consciousness, for example. Instead of supernatural explanations, however, these brain scientists sought physiological ones. Physicians Thomas Laycock and William Carpenter espoused the notion of "unconscious cerebration" that governed the relationship between the nervous system and reflexes; this concept was proposed as the theoretical basis for much of these spiritual phenomena. Somnambulism, for example, could be explained as a "suspension of volitional power."[37] A state of trance was a neurological event signifying that "cerebral activity is concentrated in some limited region of the brain, with the suspension of activity in the rest of the brain and the consequent loss of volition."[38] The medico-scientific community set out to examine the phenomena not just by philosophical reasoning about their presumed mechanism of action but also by reproducing the effects by natural means. It is presumably in this context that William James conducted his self-experiments with psychoactive drugs.

Given the Victorian propensity towards exploration of mystical happenings, it is not surprising that psychoactive drugs would hold some fascination for William James. In the 1860s as a young medical student, he experimented with drugs and had read Moreau de Tours' famous book on hashish.[39] In the 1870s, he became interested "in effects of mind-altering drugs and the possibilities of such chemicals to open new understanding of personality."[40] During the course of his self-experimentation, he tried a variety of psychoactive drugs, including chloral hydrate (the first synthetic sedative), amyl nitrite, and hashish, all "with varying success."[41] In a letter in 1870, to his brother Henry, the famous novelist, he mentions experimenting with chloral hydrate, a new sleep agent.[42] Chloral hydrate, in fact, became one of the most commonly prescribed hypnotics to treat insomnia, anxiety, and restlessness. James was plagued all of his life with insomnia and suffered from neurasthenia, a popular diagnosis during the late decades of the 19th century that was characterized by and eventually equated with nervous exhaustion.

Twenty-six years later, he was still experimenting with new drugs. In another letter to his brother, Henry James, dated June 11, 1896, he writes that Silas Weir Mitchell (the famous American neurologist and a personal friend of his) had sent him some mescal buttons to try.[43] Mescal, or mescaline as we now call it, is the psychoactive drug found in the peyote cactus and a powerful hallucinogen; the "buttons" are the pieces of the cactus that one can break off from the plant. James was keenly interested in the prospect of the intensely colored visual hallucinations that Mitchell had reported to him and would later publish that year.[44] However, the results were disappointing for James. James recounts that he wasted two days by being violently ill after consuming one bud or button. With typical wry humor found throughout his correspondence, he closes the letter to his brother with the statement "I will take the visions on trust."[45]

Subsequently, James and Mitchell's friendship would be strained and eventually ruptured. This was not due to the mescaline incident but rather because of Mitchell's open skepticism of Mrs. Piper's spiritualist abilities. He, along with other friends, had accompanied James to one of her séance's and did not react in the way that James had expected. James was insulted by the physician's open scoffing and outright dismissal of the clairvoyant claims and never met with Mitchell again, despite the latter's numerous attempts at reconciliation.[46] It seems that while James was open-minded towards a wide variety of experiences associated with human consciousness, he did not display a similar open-mindedness when in regard to contradictory opinions.

The most well-known event today related to James' varied drug experimentation had been instigated after reading an essay on "The Anesthetic Revolution" by Benjamin Paul Blood in 1874. Inspired by Blood's philosophy, James was to experiment with nitrous oxide off and on for the next couple of decades, writing about it in four works spanning almost 30 years: a postscript to "On Some Hegelisms" in 1882, an 1898 article in *Psychological Review*, his 1902 *The Varieties of Religious Experiences*, and an essay in 1910.[47] In the most well documented case, James recorded his thoughts while under the influence—purported to reveal great mystical insights; but in the cold light of sobriety, the words turned out to be gibberish. "The keynote of the experience is the tremendously exciting sense of an intense metaphysical illumination. Truth lies open to the view in depth beneath depth of almost blinding evidence. The mind sees all the logical relations of being with an apparent subtlety and instantaneity to which its normal consciousness offers no parallel; only as sobriety returns, the felling of insight fades, and one is left staring vacantly at a few disjointed words and phrases."[48]

For James, psychoactive drugs were a means to an end—another tool in the study of consciousness. "James is no dabbler, no scientific dilettante. He takes a lifetime study of the 'ultimate plurality of [conscious] states.'"[49] For him, the drugs were potential keys to unlock the mysteries of both the mind and the transcendence of that mind. Whenever they failed to provide him with the spiritual insight he craved, James discontinued the drugs.

Our third protagonist is Henry Havelock Ellis (1859–1939), best known as a pioneering sexologist who challenged Victorian taboos about sex and wrote extensively on human sexuality. He described various sexual practices and deviances and his work was considered the first objective study of homosexuality. He had become a medical doctor in order to study sex from a modern scientific standpoint but never actually practiced as a physician. He had some tangential interest in psychology and certainly in Freud's theory of psychoanalysis with its emphasis on sexual drive, and it was from that basic curiosity that led him to a brief study of psychoactive drugs. Ellis was also a promoter of empirical science in this late Victorian period of intense demarcation of new fields of study, such as psychiatry and experimental psychology. However, he was working in a period that was still a time of "open discourse, more spaciously framed in its address to common issues, and with an audience crossing wide disciplinary interest."[50] He experimented on himself with a variety of psychoactive substances, including hashish and mescaline. Like William James, Ellis' explorations in 1897 of mescaline-induced hallucinations were also inspired by reading Silas Weir Mitchell's account of taking the drug.[51] Ellis obtained a large sample of buttons from the London firm of Potter & Clarke and consumed them, one button per hour for three hours on a Good Friday, recording his responses all the while. Over the next two years, he experimented with the drug on a total of six participants, including the writers William Butler Yeats and John Addington Symonds.[52]

Ellis' investigations were first published in 1897 and 1898 in the *Lancet* and *Contemporary Review*, respectively.[53] In the *Lancet* article, he gives a more general account of visual, olfactory, and auditory hallucinations, the synesthesia experienced, the effects on cardiac and respiratory systems, and the possible use in treatment of neurasthenia. He does not go into any depth on the nature of the hallucinations.[54] In the *Contemporary Review* article, he goes into substantially more detail on the visual hallucinations produced by mescaline as he explicitly desired to add to Silas Weir Mitchell's report of mescaline from a non–American point of view.[55]

His detailed account begins with the first symptoms of faintness and dizziness and continues with a vivid description of the visual perceptions that ranged from distortions of real sensations to outright hallucinations.[56] The sensations were at first vague without form. They then transformed to geometric, kaleidoscopic images which he said constituted "the most delightful of the experience."[57] The sheer variety of images, in addition to the quantity and duration, made an impression especially since the forms were never actually known entities but at most, just semblances of actual objects. Turning on the gas lantern resulted in new distortions and shadows, analogous he thought, to the impressionist paintings that were transfixing the art world.

This latter observation inspired Ellis to recruit an artist friend to try mescal.[58] However, this self-experimenter did not have nearly the positive experience with the drug that Havelock Ellis had reported. The visual hallucinations were nonetheless equally compelling and Ellis provides another rich description of them. Interestingly, the artist wrote about the uncoupling of mind and body, to wit, a disembodied sensation of a mind observing itself and its environment from a detached viewpoint: "During the period of intoxication the connection between the normal condition of my body and my intelligence had broken—my body had become in a manner a stranger to my reason—perfectly sane and alert, for a moment sufficiently unfamiliar for me to become conscious of its individual and peculiar character."[59]

Ellis' final experiments also included himself and two poet friends as participants, one of whom he says was interested in mystical phenomena and "very familiar with various vision-producing drugs and processes."[60] Unfortunately, this friend also had a weak heart and found the cardiac effects too aversive to take much pleasure in the experience. The other friend reported a heightened sense of wellbeing and similar visual hallucinations as reported by Ellis and others. The last self-experimentation by Ellis was undertaken to examine specifically the relationship with the hallucinations to music because of comments by the previous friend while under the influence of mescal, as well as the role of music in the ceremonial administration of the drug by native Americans. During his encounter, Ellis noted the particular musical works evoked by the hallucinations. Ellis especially wanted to know if deliberate suggestion of a composer or musical piece would influence the nature of the visual hallucination. He reported that it did so in about half the instances.

Ellis was also interested in the connection between dreams and the hallucinogenic drugs, as discussed in his book *The World of Dreams* in 1911. He states the "shifting and multiplication of dream imagery ... is a fundamental and elementary character of spontaneous mental imagery, and is constant in some drug visions, notably those occasioned by mescal."[61] The similarity with hypnogogic hallucinations (those occurring during the transition of wakefulness to sleep) was also noted in his autobiography.[62]

Ellis' descriptions of the mescaline-induced visual hallucinations are markedly similar to those reported a century later by the renowned neurologist, Oliver Sacks, in *Hallucinations*.[63] Ellis' 1911 account begins with "a constant succession of self-evolving visual imagery which constantly approaches and eludes semblance of real things; in the earlier stages these images closely resemble those produced by the kaleidoscope and they change in a somewhat similar manner. Such spontaneous evolution of imagery is evidently a fundamental aptitude of the visual apparatus which many very slightly abnormal conditions may bring into prominence."[64] In comparison, Sacks' contemporary description of hallucinations associated with migraines is as follows: "Unlike the scintillating scotoma itself, which had a fixed appearance and a slow, steady rate of progression, these patterns were in continual motion, forming and re-forming, sometimes assembling themselves into more complicated forms like Turkish carpets or complex mosaics or three-dimensional shapes like tine pinecones or sea urchins."[65]

Santouse, Howard and ffytche in 2000 surveyed individuals who had visual hallucinations due to Charles Bonnet Syndrome and reported that their hallucinations tended to cluster into three categories that corresponded to hierarchical pathways of the visual system in the brain.[66] A majority of respondents reported the tessellated patterns and kaleidoscopic activity as well. These hallucinations build upon the most basic feature detectors of the primary visual cortex. These simple, complex, and hypercomplex cells respond to lines, edges, movement and from these primordial elements the brain constructs a visual image.

"In time, Ellis built up a small following of people dedicated to his experiments," but while Ellis was interested in the psychoactive drugs from a scientific perspective, his friends were equally keen on the drugs' ability to "probe the unknown."[67] After receiving an inquiry, Ellis also gave Francis Galton (a cousin of Charles Darwin and one of the early founders of the eugenic movement) a sample of three buttons to try, assuring him that there were no negative after-effects.[68] However, he did caution that one should be in good mental and physical health to experiment with the drug, particularly as mescaline resulted in a rather dramatic drop in heart rate and respiration. In a tautological conclusion, he believed that these physiological effects would render the drug harmless and not be habit-forming since only physically fit individuals could take it.[69]

While Havelock Ellis advocated the use of mescaline as a tool for psychologists to study mental states, he was cautious about promoting it for just anyone to try and in one article, he indicates mescal had no obvious therapeutic value despite previous mention of some calls for its use in treating neurasthenia.[70] He writes, "Having learned what the experiment has to teach, I have no special inclination to renew it."[71] Reflecting the attitudes of the late Victorian age, he believed that mescaline had educational value in isolating the intellect for "civilized men" in ways that had escaped the indigenous peoples who used the drug for spiritual purposes. "It may at least be claimed that for a healthy person to be once or twice admitted to the rites of mescal is not only an unforgettable delight, but an educational influence of no mean value."[72]

So here we have three physicians-turned-psychologists who published their respective accounts of experimentation with psychoactive drugs. These intellectuals were more or less contemporaries with each living in a time when new pharmaceuticals were being discovered at an astonishing pace and drug companies relied on physicians to engage in clinical trials (whether on themselves or others!) to learn about the effects, particularly potentially therapeutic, of the drugs. Zieger contextualizes their experimentation within the Victorian penchant for collecting the exotica of the far reaches of the British empire.[73] Consistent with the imperialist approach, the drugs are often used as tools for exploring inner reason of the civilized European/American, while the native "savages" use those same drugs for mindless work (e.g., coca leaves) or superstitious rituals (mescaline). Two of the three scientists, James and Ellis, pursued their respective inquiries with an intellectual curiosity consistent with their philosophical approach in studying mental life. In contrast, Freud for a time unabashedly promoted cocaine as a veritable cure-all, and only abandoned the drug work when it became publicly reviled. There is little evidence that Freud ever changed his own opinion of the drug. His only modification was a grudging admittance that the drug was not for everyone. Ellis was more cautious about mescal and remarked that "habitual consumption [of mescal] ... would be gravely injurious."[74] James advocated nitrous oxide as a means to study consciousness, simply another experimental tool.

All three were Victorian explorers rather than 20th-century modernists and were seeking to extend the boundaries of neurological science as they understood them at the time. If the nervous system was a closed energy system, then psychoactive drugs acted upon that nervous energy. They were building upon a new physiological psychology that had as its premise the unity of mental and neural events.[75] This science "was an aspect of a general movement towards a naturalistic understanding of man. Consequently, the issues raised in discussion of the relationship between the psychic and the organic were 'part of one debate; the place of mind in physical nature.'"[76] Whereas Cartesian mind-body dualism, a philosophy that had dominated science for the previous couple of centuries, had maintained that the mind could not be analyzed scientifically, late Victorian science was challenging that premise. Consciousness, according to Cartesian philosophy, was simply not "amenable to scientific study."[77] Now there were empirical methods that were acceptable to the scientific community. Just examining a few volumes from the late 19th-century intellectual journal *The Mind* reveals the range of topics thought worthy of investigation. There were articles by the leading American and British scientists, alienists, physicians, experimental psychologists, experimental physiologists, and philosophers throughout the last two decades of the century. Some articles reported on results from experimentation in sensation and perception: "The Question of Visual Perception in Germany,"

"The Muscular Perception of Space," "Harmony of Colours," and "Pleasure of Visual Form"; other articles focused on "The Definition of Instinct," "Definition of Consciousness," "The Physiological Basis of Mind," "An Empirical Theory of Free Will," and "Are We Automata?" Still other articles explored nervous disorders and disabilities: "Note Deafness," "Laura Bridgman—A Case Study" (Bridgman was blind and deaf), "Morbid Affections of Speech," and "Sense of Dizziness in Deaf-Mutes." William James, writing to the British mental philosopher James Ward in 1892, said "the real thing to aim at is a *causal* account; and I must say that appears to lie (provisionally at least) in the region of the laws as yet unknown of the connection of the mind with the body. That is *the* subject for a 'science' of psychology."[78]

In their respective experimentation with these new psychoactive agents, James, Ellis, and Freud were exploring themes important in the late 19th century as neurology, psychiatry, and psychology were emerging as distinct scientific disciplines. These themes included explorations of sensation and perception, free will or voluntary action, the relationship of theories of mind with causality.[79] For William James, experimentation—at least for a time—provided a framework to study empirically the spiritual realm. "Psychic" phenomena referred to both the emergent experimental psychology and spiritualism.

Our three protagonists did self-experiment with drugs now vilified by mainstream society—cocaine and mescaline, among others—but in a scientific, not recreational manner, as contemporary pro-recreational websites suggest. Those websites conveniently ignore the negative effects that each scientist experienced and likewise the cautions that each man expressed in regard to the drugs under study. Weir Mitchell, the American neurologist who inspired both Ellis and James to experiment with mescaline, cautioned that there would be "a perilous reign of the mescal habit when the agent becomes attainable," perhaps foreshadowing the troubled relationship with hallucinogens that future generations would have.[80] Simply extracting and publicizing the isolated facts that Freud, James, and Ellis all used psychoactive drugs as evidence to support modern day use is at best sloppy and incomplete history, and at worst, reckless exploitation of scientific reputations.

6. LSD Redux: Hallucinogens

"Lowered anxiety, demoralization, and fear of death ... improved mood and quality of meaningful interpersonal relationships." These are statements by researchers regarding their study of the efficacy of therapy with late stage, terminally-ill cancer patients. The results have now been replicated in numerous studies. But the therapy is not any of the traditional psychotherapies nor is it traditional drug therapy with antidepressants or anti-anxiety medications. The therapeutic drug used is lysergic acid diethylamide (LSD) and the decade is the 1950s.[1] For several years, LSD had shown significant results as a "psycholytic"—a drug that enhances traditional psychotherapies. But within 20 years, that research would be overshadowed by the "psychedelic" label (drugs that alter consciousness) and its reputation undermined by covert governmental research on "mind control." This chapter explores intertwining of those three threads to explain why such promising research came to a halt and how after a 30 years' quiescence, it has resurfaced.

Seeking transcendent experiences is a long-standing endeavor of the human species. Since the earliest of recorded times, individuals have sought to rise above their everyday existence through a variety of means. Many individuals, such as mystics in ancient and medieval times, have sought these experiences through extreme physical stress, such as sleep deprivation and fasting.[2] The more common avenue, however, has been through psychoactive drugs and particularly the psychedelic drugs, those that produce hallucinations: LSD, mescaline, psilocybin, MDMA ("ecstasy" or "Molly"), ketamine, and ayuhuasca, for example.

We tend to associate psychedelics, particularly LSD, as infamous products used only in the 1960s and '70s, but there has been a resurgence in recent years in illicit use of the traditional hallucinogens as well as newer synthetics, such as bath salts. More interesting is the reemergence of psychedelics for therapeutic purposes. How is it that drugs from one of the most feared and reviled pharmacological psychoactive categories are now being studied in regard to treating psychiatric disorders? As you will see, the application of hallucinogens to the treatment of varied mental illnesses is not a new concept. This chapter will trace the use and misuse of these drugs to provide a background for understanding the allure of these drugs. While the discussion is not limited to LSD, its history and usage has the best documented record in medical and psychological history. In fact, we can see a clear divide between the natural and synthetic hallucinogens in terms of historical usage. As is typical of many psychoactive drugs, the synthetic psychedelics were developed initially by pharmaceutical companies in the late 19th or early 20th centuries with therapeutic uses in mind. In contrast, the natural psychedelics have been used by indigenous peoples likely from before earliest recorded histories.

The synthetic hallucinogens include LSD, Ecstasy or Molly (MDMA or 3,4-Methylenedioxymethamphetamine), ketamine (a veterinary anesthetic), and synthetic cannabinoids, unrelated to the marijuana plant (e.g., "Spice," K2, "bath salts"). PCP (phencyclidine or "Angel Dust") is also a synthetic, originally developed as an anesthetic. These are all 20th-century products and all are currently illegal for personal recreational use in the United States.

The naturally occurring hallucinogens are numerous but the ones most commonly cited and studied scientifically include mescaline, a substance found in the buttons of the peyote cactus, and psilocybin, found in the various species within the *Psylocybe* genus of mushrooms. Ingestion of psilocybin results in the production of a primary metabolite, psilocin, which is also hallucinogenic. However, ololiqui (from morning glory seeds), ayahuasca (from the *Banistereopsis caapi* vine; also known as yagé), salvia (*Salvia divinorum*), and ibogaine (from the roots of *Tabernante iboga*, growing in the Congo) have also been researched and examined for potential therapeutic uses. Some pharmacologists classify marijuana and hashish as hallucinogens because of their ability to distort perceptions but I will not include them for discussion in this chapter. A common feature of all these naturally-occurring substances is their unpleasant and bitter taste, stemming from the alkaloid properties of their chemical make-up.[3]

There are accounts of the effects of hallucinogenic substances in ancient Sanskrit and Aztec texts, and in the works of the ancient Greeks, Homer and Herodotus. However, it was in the late 18th century that Western scientists, both amateur and professional in the burgeoning fields of botany and anthropology, began to bring samples back from exotic lands to the United States and Europe. Mescaline was the first to be systematically studied and written about in scientific journals. Louis Lewin came across the peyote cactus (*Anhalonium lewinii*) in his travels in the late 1880s. He wrote the first definitive text on hallucinogens in 1924 but had previously been responsible in collaboration with Arthur Heffter for isolating mescaline from the peyote cactus in 1896.[4] As discussed in the chapter on Victorian drug explorers, William James, Silas Weir Mitchell, and Havelock Ellis famously engaged in self-experimentation with mescaline in the 1890s. According to Bromberg and Tranter, renowned neurologist Eugen Bleuler (a contemporary of Freud's) and experimental psychologist Max Wertheimer also experimented with the drug, recommending its use to study schizophrenia.[5] The similarities of hallucinations in both the induced drug condition and the clinical condition offered an opportunity for gaining insight into the psychiatric disorder. The hope of using hallucinogenic drugs to better understand schizophrenia resurfaces periodically in the modern history of hallucinogens.

Unsurprisingly, the early self-reports on mescaline focused primarily on describing the personal experience of the user and all commented on the visual hallucinations produced by the drug. However, anthropologists were quick to note that this was not the most desirable nor dominant effect experienced by native users. They had learned with experience to moderate that effect.[6] Shultes (1938), an anthropologist who studied the use of peyote in the religious ceremonies of native Americans in the Southwestern United States, even commented that they considered peyote ill-used if the visual hallucinations were predominant, as the cactus button was also utilized in healing a variety of illnesses, such as tuberculosis and pneumonia.[7]

In the 1890s, the early scientific self-experimenters were eager to share their experiences with the medical and psychological community. They all meticulously described drug dosage effects on a variety of physiological and psychological dimensions.[8] Much of this early research was at the behest of the Parke-Davis pharmaceutical company that then supplied the peyote buttons, as evidenced by publication in their sponsored journal *Therapeutic Gazette*. William James gave it a try but found the experience not only unsatisfactory but downright aversive, becoming nauseous without any hallucinations to compensate for having the negative effects. Hoffer and Osmond, writing some 70 years later, speculated that his dose was too small.[9]

Even then, there was speculation on the similarities of altered consciousness elicited by mescaline versus what was experienced in schizophrenia. Alwin Knauer was first laboratory assistant to Emil Kraepelin, one of the "founders" of psychiatry and famous for his classification scheme for sub-types of schizophrenia. In Kraepelin's lab, Knauer and Maloney conducted a series of 23 mescaline experiments upon themselves and other individuals associated with Kraepelin's lab.[10] These experiments were perhaps the first systematic attempt to examine different dosages with a different route of administration, i.e., sub-cutaneous, which meant nausea was minimized, rather than the usual oral route. The visual hallucinations described in these early reports were consistent in their kaleidoscopic patterns, intensification of color, and perceptual distortions. Kraepelin, one of the first experimental psychologists in addition to being a psychiatrist, believed that mescaline could be used to study schizophrenia as well.[11]

Despite Ellis' conclusion that mescaline would not add anything to the "already overcrowded field of therapeutic agents," the 1920s brought more rigorous testing methods to examine mescaline's effects on those possibilities.[12] Klüver's 1928 monograph and Beringer's 1927 work *Das Meskalinrausch* are considered the classic texts on mescaline and the detailed descriptions provided therein have not been surpassed in their thoroughness.[13] "Mescaline made it possible for the first time to investigate the phenomenon of hallucinations from a scientific, pharmacological, and clinical aspect using a pure chemical compound."[14] Psychopharmacology was beginning to take off as advances in analytical chemistry brought forth new substances with therapeutic promise for psychiatric disorders.

The research environment was thus ripe for the discovery of LSD. Albert Hofmann, a Swiss chemist working for Sandoz pharmaceutical company (precursor to the pharmaceutical giant Novartis), is credited with its discovery. While often reported now as a chance discovery, Hofmann stated that it was more along the lines of what we now would call "planned happenstance."[15] Hofmann was tasked by Sandoz in 1938 to synthesize and study all the alkaloids of lysergic acid, one of the active compounds found in the ergot fungus. The putative reason at the time was the hope that one of the derivatives would be useful in stopping uterine hemorrhaging during childbirth because of ergot's effect on oxytocin, a natural hormone which was known to induce uterine contractions. Hofmann expected the diethylamide to act as a stimulant and preliminary animal testing on LSD-25 (the 25th compound in the amide series to be synthesized) indicated some excitation in the animal subjects. However, not much of interest was noted in his laboratory experiments and he put aside the compound.[16]

Five years later in April 1943, Hofmann decided to take another look at the compound for some further study. This trial period was the source of the serendipity legend of LSD. Apparently, he accidentally ingested some of the compound without being aware he had done so. He began to feel dizzy and restless to the point where he had to go home in the middle of the workday. Subsequently, he experienced a sensation of drunkenness and began to hallucinate vivid images "in a kaleidoscope-like play of colors."[17] His symptoms were not consistent with ergot poisoning yet he reasoned that the experience had to be due to the compound he had been working with in the lab. The greatest puzzle to him was how such an amount, minute enough to be unnoticed by him, would produce such dramatic effects!

He therefore decided to engage in more systematic self-experimentation while he tried the lowest possible oral dose of 0.25 mg.[18] He kept a journal log to record his reactions but was thwarted because after an hour, his notes were indecipherable! Then with the aid of a laboratory assistant, he tried again. From that set of self-experiments, he ascertained that an effective dose was .03–.06 mg. In comparison with mescaline, LSD was "5000–10,000 times more active and produced qualitatively nearly the same symptoms."[19] LSD was unique in its potency but similar in effects to other known hallucinogens at the time: mescaline, teonanacatl (foliate mushroom) and ololiuqui (related to the flowering plant, morning glory).

Thus began the post-modern scientific research on hallucinogens that was to last almost 25 years before legislation effectively ground it to a halt. In 1949 Sandoz supplied LSD (known by its trade name Delysid) to researchers.[20] By 1951, more than 100 research articles had been published in scientific journals.[21] In another ten years, the number of research reports had grown by over ten times. This surge was a function of the rise of biological psychiatry and psychopharmacology in part inspired by Hofmann's key discovery but also catalyzed by an explosion of research on psychoactive drugs and their effects on neurotransmitters in the brain. The finding that an infinitesimal amount of a drug could produce symptoms similar to those in a serious mental disorder, schizophrenia, changed the research landscape. The research in the early part of the century yielded rich descriptive data; now the focus was turned to application. How might LSD be used in a clinical capacity?

Sidney Cohen, one of the leading LSD researchers during those post-war decades wrote the following in 1970 as that research era was coming to a close:

> [The significance of Hofmann's discovery] demonstrates that chemical substances in extremely minute amounts can induce mental distortions that resemble the naturally occurring psychoses; it has stimulated interest in the chemistry of the nervous system, especially the chemical transmitters across the synapse, the nerve-cell connections; and it permits the laboratory study of both normal and abnormal mental processes—an approach to the understanding of such phenomena as attention, imagination and perception as well as hallucinations, delusions and depersonalization becomes possible.[22]

Put another way, the LSD discovery revealed a possible explanation for certain mental disorders, particularly schizophrenia. It was now seemed plausible for a previously undetectable amount of a chemical in the body to cause cognitive changes and dramatic symptoms, such as hallucinations.[23] This was considered a landmark discovery in chemistry at the time and helped precipitate the biochemical research that ultimately led to the discovery of neurotransmitters.

Research on therapeutic usage in the following years tended to concentrate on one of two central research questions: First, could LSD or other hallucinogens provide a research model of psychosis? Second, could they be used to treat specific psychiatric disorders? The first question was problematic from the start. Obviously, inducing a psychotic episode in normal participants for the purpose of modeling psychoses posed an ethical challenge. Researchers believed the effects of the drug would wear off as the drug itself was fully metabolized by the body and so presumably, the abnormal condition was reversible. However, experiences varied widely among individuals, being influenced by environment, past personal history, and participant expectations. Besides the ethical concerns, research designs were fraught with uncontrolled factors. These problems were not unique to hallucinogen research but would later provide the impetus for the cessation of scientific studies.

In order to address some of the issues stated above, researchers turned to animal models of specific mental illnesses to study. Focusing on a biochemical approach with the new and exciting work on neurotransmitters was a natural path to follow now that there were drugs that seemingly mimicked the course of certain mental disorders. The advantages of using animal models include far greater experimenter control over the animal subjects' environment and genetic endowment.

But the notion of using animal subjects for experimenting with drugs with such unique effects as hallucinations raises created its own set of new challenges. How can the distinctive hallucinogenic experiences that are cognitively impenetrable by an observer be effectively studied in animals? One such attempt used spiders as subjects. Peter Witt, in a series of experiments beginning in 1948, famously gave spiders a variety of psychoactive substances and found that their subsequent web-making abilities were altered in distinct ways. Interestingly, the LSD web was more intricate, while the mescaline-induced web was disorganized.[24] The studies provided an objective measure to differentiate effects among the separate drugs and certainly produced fascinating results but from a practical standpoint, the mental life of spiders is not comparable to that of humans when looking for treatments of mental disorders. Hallucinations, a hallmark symptom of schizophrenia, create a uniquely human, subjective experience and no animal model would provide a satisfactory approach to study the illness.

Osmond and Smythies in 1952 first proposed a new mescaline model to study schizophrenia based on the hormone epinephrine (also known as adrenaline).[25] That "adrenochrome" theory was advocated by Smythies for many decades. While their specific premise that adrenaline malfunction underlies the acute schizophrenic state was overtaken by experimental focus on other neurotransmitters (e.g., dopamine and glutamate), as well as the role of genetics, the key point for the present discussion is that it was perhaps the first drug model of a mental illness. This represented one of the prongs of psychiatric research with hallucinogens.

Despite research developments indicating the role of chemical neurotransmitters in brain function and thus underlying normal and abnormal behavior, ultimately, researchers backed away from the idea that LSD or mescaline in humans actually mimicked schizophrenia.[26] For example, an early study of mescaline with a sample of psychiatric patients with varied disorders indicated that the drug not only had different effects on individuals with different disorders, but it also had effects on discrete symptoms and not the overall pattern of psychosis.[27]

However, as the idea of LSD-induced psychosis as a model for schizophrenia waned, the notion that LSD could be used as an adjunct to psychotherapy in the treatment of other psychiatric disorders began to be vigorously pursued. Starting in the early 1950s, this combination was studied most extensively with alcoholism in a major research program by Osmond and Hoffer in Saskatchewan, Canada. The original intention was to look at a biochemical model of mental illness with the "scientific observations of a subjective experience."[28] They believed their model reflected a new approach in blending psychopharmacology with psychotherapy traditions. While late 19th-century remedies for mental disorders and drug experiments had been anchored in physiological nervous system theory, Freudian psychoanalysis had dominated American psychiatry from the 1920s through the 1960s. Although Freud had envisioned a neurological basis for his theory, that original intention was by then largely overlooked in part due to lack of evidence. Even today in my teaching, I find students to be mostly unaware that Freud's theory was meant to be a biological theory. The practice of psychoanalysis over time has obscured that origin for the public understanding although there is a bit of a resurgence in applying contemporary neuroscience to Freudian concepts.[29]

The aftermath of World War II created a need not only for more therapists to deal with the needs of returning soldiers but a climate for development of faster, more efficient therapies than psychoanalysis. Group therapy was one such solution but there was renewed interest in somatic therapies and, of course, pharmacotherapies. So, given that psychiatric environment, the notion of both nature and nurture interacting to result in mental disorders was not as well accepted by scientists as it is today and thus the Canadian research program was pushing the frontiers of psycholytic research.

Researchers in 1953 had found that LSD in psychiatric patients intensified emotion and memory, leading to significant clinical gains in therapy.[30] Within a few years, clinicians were combining classical psychoanalysis with low doses of LSD. Note the LSD was an adjunct to therapy, not the sole therapy. Between 1953 and 1968, more than 7000 patients were treated with this combination and there were 18 special treatment centers in Europe alone.[31] Kast in the 1960s conducted now famous studies with terminally-ill cancer patients and results indicated that LSD eliminated ego boundaries, increased depersonalization, and improved psychological well-being.[32] Other psycholytic studies showed improvements in patients with anxiety, depression, and alcohol addiction.[33]

Studies on using hallucinogenic drugs with alcoholics made up the largest component of the new research program.[34] Anecdotal evidence from Native Americans suggested that those who had alcohol abuse problems who had turned to the peyote cult had given up alcohol successfully. Researchers Osmond and Hoffer in Saskatchewan were struck by the similarity of LSD-induced hallucinations and the delirium tremens (DTs) in alcohol withdrawal. Because alcoholism was viewed according to the lens of a "moral" model where the condition represented a loss of personal control, LSD with its effect on depersonalization and fragmentation of ego boundaries might seem an odd choice of treatment drug. However, the drug was not intended ever to be a treatment or cure by itself; its role was to put the patient in a state that would provide less resistance to the traditional talk therapies. Results were promising but initially faced questions of credibility.[35] The evidence that participants stopped drinking as a result of the combination therapy seemed too good to be true, and in fact, replications yielded mixed results perhaps due to the sensitivity of LSD effects to different environments and protocols.

Hoffer and Osmond's work received more favorable attention as alcoholism began to be reconceptualized as a disease rather than a signature failure of will power. "The medicalisation of alcoholism expanded clinical authority into an area governed by political and social decisions."[36] Hoffer and Osmond advocated an approach that relied on a single "mega-dose" of LSD to short-circuit attentional processes, followed by talk therapy. Treatments that did not incorporate psychotherapy with the use of hallucinogenics yielded negative results.[37] According to Hoffer, studies of more than 700 alcoholic participants showed a 50 percent recidivism rate.[38] That rate may not sound all that impressive but no other treatment at the time came close to producing those results.

Proponents of the psycholytic therapy maintained that LSD was effective for a variety of compelling reasons.[39] Using the psychoanalytic framework of the time, they said the drug lowered defenses and allowed repressed psychic material to emerge into consciousness; because the material was visual, as portrayed in the visual hallucinations, it was better understood by the patients; the psychic material was experienced without the crippling effect of guilt; the drug experience brought together the patient and therapist into a stronger, more intimate relationship of trust conducive for psychotherapy; and finally, and perhaps most importantly for therapy, the insights gained during the experience were retained afterwards. LSD used in the therapeutic environment was also quite safe.[40] Few adverse effects were seen, nor was there any indication of tolerance or addictive qualities. Yet neither was there clear consensus on how the drug worked and how best to use in a therapeutic environment.

But in the 1960s, a shift in emphasis from psycholytic to psychedelic treatment occurred. Whereas psycholytic approaches dealt with the underlying biochemical basis of both the drug and disease, psychedelic treatments focused on the transcendental experiences that many individuals had from taking hallucinogenic drugs. Osmond had coined the term "psychedelic" during communications with Aldous Huxley, the famous writer and philosopher who would become one of the outspoken champions and public promoters of LSD.[41]

The proverbial cat was now out of the bag. As the public became more aware of LSD, recreational use increased. According to Dyck (2005), participants for the medical studies were volunteering more for entertainment purposes to get high (although one could question why any "normal" individual volunteers for drug studies—usually the answer is compensation).[42] Therapists with little training began to use LSD in their sessions but without the careful oversight and preparation that the pioneering clinicians had used. Therapists were also using LSD and other psychedelic drugs indiscriminately among their psychiatric patients despite considerable research that indicated that those drugs were not effective for some disorders and in some cases, created adverse effects.[43] Some researchers have since seen some of these critiques as reflecting turf wars in academe: who had the authority to say how LSD and other hallucinogenic drugs should be investigated.[44] As LSD became more readily available to the public via the drug scene, problems began to appear with contamination in combinations with other psychoactive drugs such as alcohol, cannabis, barbiturates and amphetamines.

And then there was the Harvard group. The names Timothy Leary, Aldous Huxley, and Allen Ginsberg are eponymous with the psychedelic experiences that infiltrated the cultural upheaval of the 1960s. Psychologists Leary and Richard Alpert first were involved in the psilocybin project at Harvard that eventually drew in writers Huxley and Ginsberg. This was a research project to examine effects of psilocybin and its potential use in psychotherapy. Leary's goal was to show that not only could psilocybin induce a transcendental experience with enhanced self-understanding but would result in positive behavior change as well. Over time it became further and further removed from any scientific research paradigm and enjoyed an almost cult-like status at Harvard. By the time Harvard administrators became concerned about the questionable reputation of the project and its directors, with reports of indiscriminate drug use and drug-fueled orgies, including the "Good Friday" experiment, the proponents had already discarded psilocybin in favor of a newer drug: LSD.[45]

Even before Harvard finally acted to fire Leary and Alpert, they had already largely abandoned any pretext of social science research on LSD and instead created a more public campaign to promote LSD as a mind-enhancing and mind-expanding drug for the masses.[46] More interested in contributing to the growing social movements, in 1963 Leary had created the IFIF, the International Foundation for Internal Freedom. This was to be a loosely connected body of independent groups dedicated to the exploration of LSD and its mind-altering effects. The Federal Drug Administration

(FDA) had by this time gained control over all experimental drugs with the passage of a 1962 law. In order to maintain a supply of LSD from Sandoz, the IFIF would still have to engage in legitimate scientific research. The IFIF goal was to have a critical mass of the American population using LSD to result in psychical revolution for societal change.[47] However, IFIF had a short life and was defunct by the end of 1963. Leary and his fellow psychedelic enthusiasts took up residence at the New York house, Millbrook, where they continued to promote LSD. The whole affair effectively came to an end with a raid in 1966 that shut down Millbrook.

LSD's reputation had done a complete turnaround by the mid–1960s from promising adjunct to psychotherapy to one of the most reviled psychoactive substances to hit America's youth scene. Whereas Sidney Cohen had shown in his research in the 1950s that LSD was remarkably safe to use within the carefully controlled clinical environment, reports of adverse effects mounted.[48] The scientific work of the preceding decades was overshadowed by the sensationalist accounts of the negative effects of LSD in the general public. According to Stevens, the lack of consensus in the scientific community for how LSD achieved its effects contributed to the snowballing of LSD as a potential therapeutic agent.[49] Government officials worried that LSD produced nonconformists and represented a threat to American values.[50] At the same time, the CIA was engaged in their own experimentation with LSD. Sidney Cohen, considered one of the voices of reason among LSD proponents, blamed Leary and his followers for unleashing LSD indiscriminately in such a way as to result in its banned status and removal of financial support from any research funding agencies.[51] The dramatic increase in recreational use among the public "doomed legitimate research by tarnishing the drug's reputation, discouraging researchers from using it, and criminalizing the use of the drug, making it harder to gain approval for research."[52] Cohen had also decried the increasingly sloppy practice by some psychiatrists who were nominally engaged in "research" in order to gain access to the drug from Sandoz but who were dispensing it without the previous therapeutic safeguards or care towards what types of patients were being given LSD.[53]

While most accounts of the cessation of LSD research focus on the rise and irresponsible promotion for recreational use of the psychedelic drug by '60s icons, Leary, Ginsberg, and others ultimately leading to the criminalization of LSD and classification as a Schedule I drug, there were at least two other contributing factors. While a thorough account is outside the scope of this book, one such factor came from the initially covert experimentation on hallucinogens by the CIA. This previously classified project came to light during Senate hearings led by the late senator Edward (Ted) Kennedy in 1977. While the testimony by CIA officials, including the director of the project, MK-ULTRA was arguably vague and misleading, the conclusion of the Hearings was to shut the proverbial door on the investigation. They declared that the 25 years of experimentation had been justified in the context of Cold War fears about communist enemies of the state engaged in similar experiments to use against the American people. Furthermore, those experiments had now ceased.[54]

In fact, the CIA and branches of the military conducted experiments on every known hallucinogenic and common psychoactive drugs available during the post-war years through the 1960s. There was intense interest in developing chemical weapons that would be "incapacitating agents," producing temporary effects rather than outright death or permanent disability in its victims.[55] LSD, mescaline and psilocybin were studied precisely for their temporary "psychosis-producing" effects.[56] LSD was an attractive option because it was inexpensive to manufacture and the minuscule amount needed for an effective dose meant it could be delivered surreptitiously and potentially in mass quantities, such as via a municipal water supply. While notions like these might seem more at home in the pages of tabloid news, these were indeed serious discussions in the U.S. Department of Defense.[57]

A second avenue of military research was the use of hallucinogens and other substances as drugs of interrogation.[58] The government had been interested in developing a "truth serum" since the early experimentation with cannabis in the early 1940s. That interest was further developed when the post-war reports of how the Nazi SS had used mescaline for interrogation purposes became available.[59] The CIA became aware of LSD's potential in the early 1950s and initially believed it to be the ideal truth serum. However, as research progressed, results showed that LSD was not as straightforward in its effects as first believed, nor did it result in amnesia of events during the drug state.[60] More troubling was discussion of using the drugs to counter "Ghandian methods" of nonviolent resistors during the civil rights era.[61] However, the same properties of LSD that made it an attractive adjunct in psychotherapy because of the breaking down of psychological defenses posed potential value for the government in rooting out counter-espionage agents. Yet it was the wide variation in individual response to LSD that also made it problematic both as an offensive weapon and as an interrogation tool.[62] While there was more or less consensus that LSD acted to reduce inhibitions, those inhibitions might be what normally controlled aggressive behaviors, for example. Effects on individuals varied according to personality, expectations, environment, and emotional health status. Critics pointed out that results of laboratory experiments could not be generalized to real combat situations and that dosage control was virtually impossible with mass distribution. While in carefully controlled laboratory settings, adverse effects with LSD were rare, there was no such guarantee against long-term adverse effects in victims of a covert attack. There were some critiques on ethical grounds: was this approach humane? How would the practice be reconciled with the American value and rhetoric of individual political freedoms?

Notably absent from these early critiques were questions concerning the ethics of the research studies performed by the CIA. Covert LSD experiments were performed without informed consent on drug addicts, prisoners, minorities, the mentally ill, and other groups.[63] Participants were not in a position to decline and sometimes completely unaware of what had been done to them; nor was the public aware of these experiments. Additionally, scientific research with LSD in universities and hospitals was carefully monitored by the CIA. The CIA then secretly worked with an American pharmaceutical company, Eli Lilly, to synthesize LSD so that the government would not be dependent on a foreign company, Sandoz, for its supply. Irresponsible use within the CIA among employees led to at least one suicide but still the experimentation continued. It expanded to the "field" with safe houses being established in Greenwich Village in New York and San Francisco. Here individuals unknowingly were given LSD, often slipped into their drinks. These practices did engender internal debates within the CIA on the ethics involved but ultimately cold-war fears won the argument.[64]

The changes in legal status of LSD in the 1960s was inconsequential for the CIA as a provision in the new law allowed for governmental use. Finally, the CIA abandoned its LSD research program, not because of ethical considerations but because the hallucinogenic was proving to be unrealistic to use in any kind of controlled fashion.[65] In psychology, the behaviorist paradigm was firmly entrenched with methods to predict and control behavior. A drug that resulted in altering consciousness with transcendental overtones was increasingly a less attractive option. However, the public disclosures in the 1970s of the government research likely contributed to the negative reputation that LSD developed and the waning interest in future research on the hallucinogen.

The other thread in this tale of how LSD research came to an end 50 years ago comes from a much more prosaic factor: the gold standard in pharmaceutical research of the double-blind experiment. While the criminalization of LSD certainly resulted in much of the research drying up, it did not mean a comprehensive prohibition of research. It was simply more difficult legally to get the drug to study and by 1970, one had to obtain permission from the Bureau of Narcotics and Dangerous Drugs. Some research on LSD continued through the 1970s. A major stumbling block proved to be one of the stipulations of the Drug Amendments of 1962, known as the Kefauver-Harris Amendments.[66] This legislation, spurred by the devastating effects of thalidomide, mandated that a clinical drug's approval was contingent on demonstrating the drug's efficacy on its proposed purpose using accepted scientific methods of the time. In effect, the law enshrined the benchmark of placebo-controlled clinical trials with double-blind, randomized assignment for FDA approval as a therapeutic agent. In other words, matched participants had to be randomly assigned to a treatment group or a placebo group with the experimenters and the participants themselves unaware of which group the participant was in during the study. This allows researchers to examine the efficacy of a drug relative to an inactive substance without either experimenter bias or participants' outcomes being affected by their own expectations.

The problem of using a double-blind, placebo-controlled approach with LSD or any hallucinogen is readily apparent. First, the hallucinogenic effects of LSD are so profound as to deter any semblance of unawareness of both participant and research observer as to who has taken the active substance. The double-blind component was simply unrealistic to maintain. The hallucinations themselves were the desirable therapeutic catalyst for successful psychotherapy with their resultant dissociation and breakdown of psychological defenses and so how could there be a comparable placebo? "Its efficacy lay in the psychological impact of the subjective drug experience, which was crafted through a unique relationship between the patient, therapist, and drug."[67] There were simply too many psychological confounds to determine conclusively that beneficial effects were due specifically to the use of LSD.

Second, the method presumes readily assessable outcomes, typically physiological in nature, in order to compare drug versus placebo.[68] However, when the outcome is psychological, dependent upon not only introspection and self-report of subjective transcendental experiences but the relationship with the therapist, that comparison lacks validity. Furthermore, LSD was not meant to be used as drug therapy by itself but as an adjunct to psychotherapy—LSD had no inherent value as a therapeutic drug by itself. Before 1966, comparisons tended to be between groups with standard psychotherapy and psychotherapy plus LSD, an acceptable research paradigm during those years.[69] While the FDA did not specify the methodology to be used for approval, according to Oram, the administrators in charge of the approval process were very much adherents of the clinical trial approach.[70] The few studies done under FDA aegis showed mostly negative results. The fact that the methodologies in those negative studies strayed significantly from what had become standard protocol from the past 15–20 years was overlooked. The changes in approval process and lack of research funding ultimately sealed the coffin on LSD research and in 1974, LSD was proclaimed by the FDA to have no therapeutic value.[71]

That should have been the end of the story except that LSD and other hallucinogens tend to be rediscovered every generation. Dyck makes a case for an approximate 30-year cycle between research spurts: Weir Mitchell, William James, and Havelock Ellis in the late 1890s, Hofmann, Klüver and Beringer in the 1930s, followed by Osmond, Hoffer and Cohen in the 1960s.[72] The appearance of MDMA or Ecstasy in the 1990s brought the next resurgence of interest in LSD research. At that point, LSD had been around both in licit and illicit forms for 50 years. The Swiss put together a symposium to examine both the history and potential future of LSD in research and therapy.[73] Several symposium participants decried the virtual cessation of LSD research in the United States and most of Europe and called for a renewed effort to convince government funding agencies to consider a reexamination of the efficacy of what had seemed to be such a promising drug.[74] Private funders had founded the Multidisciplinary Association for Psychedelic Studies (MAPS) and the Heffter Research Institute, so some research had continued in Europe.[75] By the time the proceedings of the symposium had gone to press, the FDA had loosened some of the restrictions.[76]

In the intervening years up to the present day, research on LSD has proceeded cautiously, building on the foundation laid 50 years ago or more but applying new methodologies to explore some of the same areas of promise.[77] LSD and other hallucinogens, such as psilocybin, ketamine, MDMA, and ayahuasca, are currently being studied in clinical populations in psychiatry and oncology. Researchers have focused on using psychedelic drugs to treat intractable depression, anxiety, obsessive-compulsive disorder, post-traumatic stress disorder, and addictions.[78] Besides incorporating better experimental designs with matched controls, these studies are also using statistical procedures that better evaluate small sample sizes.

Another key difference for studies now versus then is the availability of more neuroscientific information regarding how hallucinogens affect brain function and relating that knowledge to what we know about the pathophysiology of various mental disorders. The psychopharmacological work on the neurotransmitter serotonin (5-hydroxytryptamine or 5-HT) since its discovery in 1949 has revealed its important role in both mood disorders and LSD action.[79] Research established the existence of multiple serotonergic receptor sites and pathways which has helped elucidate mechanisms and brain structures involved with depression, in particular. Contemporary drug therapy using antidepressants relies very heavily on this research. The early discovery that serotonin was responsible for the psychedelic effects of LSD helped reveal the potential therapeutic link and specific receptors have been identified.[80]

During the last 20 years, neuroscience research has developed to the point that imaging technologies can now be used to study the effects of hallucinogens in the brain. For example, PET scans have been employed to examine psilocybin as a model for psychosis.[81] Robin Carhart-Harris and colleagues have conducted the first fMRI studies of individuals under the influence of psilocybin and LSD, respectively.[82] From these studies, physiological evidence is mounting that confirms the reports from researchers 50 years ago that the therapeutic merit of these drugs may lay in their ability to disrupt functional connectivity, in this case, neural connections. Enthusiasts claim dysfunctional patterns that have been intractable now become dissociated under the influence of LSD, for example, providing the neurological and psychological substrate for new healthy connections to be made.

Here the LSD story differs from the other topics in this book in that the current resurgence of scientific interest in the drug has simply picked up where it was left off in the early 1970s after an almost 50-year hiatus. There has been no dramatic change in theory driving the current research; rather, there has been a relaxation of constraints imposed more due to societal changes and popular abuse than any obvious scientific evidence of adverse outcomes from LSD. When LSD appeared in the public forum, used indiscriminately without professional oversight, the negative outcomes began to mount. Currently, while illicit hallucinogens continue to remain available in the public forum, its usage is greatly diminished compared to the 1960s and '70s. One can only wonder if that will continue to be the case if hallucinogens are eventually approved for therapeutic use.

7. The Poppy's Curse: Opiates

IMAGINE THE DAILY SUFFERING of a patient whom I will call Ms. O. She is middle-aged, middle-class, and has been a stay-at-home mom. She awakes each morning feeling just as tired as she did when she retired the previous night. Her dull lethargy is accompanied by little appetite. She feels a bit shaky and her hand even exhibits a slight tremor as she reaches for her morning coffee. A tentative sip results in vague and disturbing sensations in her bowels and she breaks into a clammy sweat despite the coolness of the room. Worse, as soon as she awoke, she is beset by the chronic, unrelenting pain—some radiating from her lower spine down her left leg but also accompanied by an overall, dull pain that seems to entrap her whole body. Her temple throbs with what will soon be a full-blown headache. The medicine, prescribed by her physician for her pain, is at her bedside and she quickly takes her first dose of the day. With that description, Ms. O could just as easily be a patient in 2018 or 1875. The essential drug would be the same: opium. "Every drug 'epidemic' hits the headlines as if it were a new and unique threat."[1] As this chapter will show, the opiate epidemic is hardly a new phenomenon.

According to the Centers for Disease Control (CDC), there is currently a drug epidemic in the United States and it's largely due to legally prescribed medication.[2] A *Washington Post* headline states, "Deaths from Opioid Overdoses Set a Record in 2015."[3] The drug or, more accurately, class of drugs, is prescription pain killers—opiates. These are drugs originally derived from opium, which is the dried sap from the poppy, *Papaver somniferum*. There are currently no existing drugs in a physician's pharmacological arsenal that are as effective in the relief of pain, especially for individuals suffering from chronic, debilitating pain. Commonly prescribed drugs in this category include morphine, Percocet, Oxycontin, oxycodone, vicodin, hydrocodone, Darvocet, and Fentanyl. But use of these drugs comes with a price: increased risk of overdose as the same drug that dampens pain perception also diminishes brain activation and ultimately depresses respiration. The recent alarm raised by the CDC concerning these drugs has been not only in regard to the sheer volume of prescriptions written for opiate painkillers but the attendant number of deaths by overdose, which almost tripled between 2002 and 2015.[4] Recent statistics indicate that the number may have seen its peak since alarms were sounded by public health officials earlier in the decade. Unfortunately, the recent decline has been accompanied by a rise in heroin and fentanyl (a synthetic opioid) overdoses and this is no coincidence. Deaths due to heroin increased six-fold between 2002 and 2015.[5] Fatal overdoses due to fentanyl have increased from approximately 3100 in 2013 to 20,000 in 2016 in the United States.[6] Since 2009, death by drug overdose, the majority of which was caused by opiates, has consistently exceeded the number of deaths by motor vehicles. For white adults, mortality rates due to overdoses have increased sharply, most notably among 25–34-year-olds and among females.[7] For the first time in more than 50 years, life expectancy in the United States has declined two years in a row and the cause is suspected to be opiate overdoses.[8]

The number of opiate prescriptions is noteworthy in its own right. Prescriptions increased by 600 percent during the ten-year span of 1997 to 2007 and continued to increase up to 2013 until leveling off in response to increased public awareness of the opiate epidemic and new CDC prescription guidelines and state regulations.[9] Equally alarming is the reported increase in use of these drugs for recreational purposes rather than therapeutic purposes, especially among young teens. For that demographic group, prescription opiates have been the fastest growing category of abused drugs, although there is some evidence that levels may have peaked.[10] One local county has seen the related number of fatal heroin overdoses among adolescents double in the last decade.[11]

So what accounts for this dramatic rise in prescription pain killers and what are contributing factors? How was this then responsible for the resurgence of heroin, not among the stereotypical drug-ridden ghettoes of the large cities but among suburban and even rural populations?

A key factor arises from the shift in medical philosophy concerning pain management. Within the medical community, proponents of the fields of palliative medicine and pain management have advocated strongly the need for appropriate use of pain medication to treat chronic pain, especially for those with terminal illnesses. Twenty years ago, pain management therapies were lacking both in sophistication and resources and there was a reluctance to prescribe opiate class drugs. This was a carry-over from the earlier debates surrounding the Harrison Drug Act of 1914. The drug panic that led to the passage of prohibitive legislation created what would become an entrenched belief of the enslaving nature of opiates.

What changed was first rise of the hospice movement in the 1960s and, secondly, the publication of an article by Portenoy and Foley in 1986 in *The New England Journal of Medicine* that contained among its references a 1980 "letter to the editor" by Porter and Jick, which stated that only a small percentage (specifically, one percent) of terminally ill patients actually became addicted to opiates.[12] This assertion was picked up in earnest by Cecily Saunders, founder of the hospice movement in the United Kingdom, and used initially to advocate for palliative care for the terminally ill cancer patient. Later, the discussion and application broadened to include any patient with chronic pain. As both hospice care and palliative medicine became more mainstream, many physicians, representatives of medical organizations, and researchers in the field bemoaned the ultra-conservative approach that needlessly deprived legitimate pain sufferers from proven relief for fear of causing addiction. The World Health Organization had developed guidelines in the 1980s for treating the terminally ill and opiates were a critical part of the care plan.[13] Pain Medicine was established as a legitimate medical subspecialty and pharmaceutical companies jumped on the bandwagon. And so, as an effort to re-educate practitioners was made, pain assessment was made the "fifth vital sign," and prescriptions began to rise. Within ten years from the publication of the Portenoy and Foley article, "a revolution in medical thought and practice was underway."[14] Aggressive marketing by pharmaceutical companies and advocacy by physicians who had financial conflicts of interest with the industry also complicated the picture.[15] In fact, Foreman maintains that the United States simultaneously has two public health epidemics: undertreated pain and opiate abuse.[16] This leads to what some call the "opioid conundrum"—it is easier for an addict to get opiates illegally than a patient to get legitimate medical relief from chronic pain.[17]

With the increase of prescriptions and parallel rise in the use of methadone to treat heroin addictions in some states, so-called "pill mills" arose to exploit the developing chronic demand for prescription pain killers. Armed with doctors and pharmacists more keen on money than sense, opiate prescriptions were written and dispensed with little oversight and with an aim for profiteering.[18] During 2014, many states, most noticeably Florida, began efforts to shut down illegal operations and enacted legislation to crack down on these pill mills and prevent "doctor shopping" by users. While this may ultimately decrease the inappropriate prescribing practices, these changes do not address the addiction problem itself: the demand that drives the supply. Predictably, in states such as Florida and Tennessee where such legislative efforts have been successful in stemming the over-prescription of opiate analgesics, there has now been a subsequent rise in heroin use as opiate addicts search for other sources.

Accessibility is likewise a contributing factor. The ready availability of opiate prescriptions via pill mills has just been discussed but perhaps more disturbing is the accessibility of prescription pain killers to teens in their own households. As more therapeutic prescriptions are written nationally, more opiates are in home medicine cabinets. First, there has been an increase in back pain and sports-related injuries, with the latter especially among children and adolescents. This in turn leads to a concomitant increase in prescription pain killers. Studies show that adolescents are more likely to obtain their opiates from their homes than on the street.[19] Sometimes it is their own prescriptions from a sports or other accidental injury but often it is a family member's prescription. Similarly, the demographic most at risk from opiate addiction in the workplace are health care professionals—again the key is accessibility.[20]

Prescription opiate addiction is insidious: first, prescription opiates are considered "medicine." Usage typically begins as a therapeutic measure, prescribed by a professional, licensed health care provider to alleviate pain caused by a physiological condition. There is a generalized public perception of prescription medications that these drugs are safe precisely because they are dispensed by a professional. The public mostly has trust in the safety system, the Food and Drug Administration (FDA), the federal organization that regulates drugs, and while it would be hard not to be aware of the dangers of opiate abuse, given celebrity overdoses and media reporting, most patients who are in pain are going to take the relief given; second, we live in a culture of quick fixes and Americans especially like to fix their problems with a pill—far less effort required than physical therapy and permanent lifestyle changes. Because of the dozens of pain killers available by prescription, patients might sometimes be unaware that the drug they are taking is in the opiate class. Health psychology shows us the power of individual patient health narratives: how one conceptualizes an illness determines the validity and efficacy of the treatments offered. Similarly, the pharmaceutical industry has encouraged us through direct marketing to ask for medications to soothe all our ills. Much has been written on the molding by the drug companies themselves of patient and physician attitudes and beliefs concerning pharmaceuticals.[21]

A critical question in examining the problem of addiction is best exemplified by Musto in the now classic 1999 work *The American Disease*: "Can anyone become addicted or are only certain persons, psychopaths or biochemical misfits, likely to succumb?"[22] Is addiction some inherent property of the drug itself, implying that anyone using the particular drug can potentially become dependent? That is the underlying assumption when we use the phrase "addictive drugs." Alternatively, is addiction a function of the person? Do only some people get addicted? Is there an addictive personality or an addiction gene? As you will see, theories of addiction have struggled with these questions both now and in the past.

Contemporary theories of addiction fit into one of four basic models: Moral, Disease, Physical Dependence, and Positive Incentive models.[23] While these models initially appear to be distinct from each other, the reality is much more blurred and some of the distinctions are revealed to be superficial. In fact, these models co-exist with one another and are not necessarily mutually exclusive.

The Moral model has held sway off and on for decades, if not centuries, and is bound in philosophical and religious rhetoric. This model holds that addiction is a function of weak will power—a character flaw, whether due to personality or one's genetic endowment. The "Just Say No" and DARE drug use prevention programs reflect this thinking. One merely has to resist temptation to stay out of addiction's web. This conservative view has enjoyed periodic support by those who favor personality typologies despite the failure to provide much evidence for a simple "addictive" personality. As you will see in the chapter on typologies, rarely is a dramatic reductionist approach satisfactory in explaining the complexity of human personality.

Perhaps the more compelling evidence for a dispositional tendency towards weak will power comes from Mischel's "marshmallow test." Walter Mischel, a prominent cognitive psychologist, has been able to show that the results of a simple test of delayed gratification in young children can powerfully predict adult outcomes related to achievement.[24] The classic test situation is to seat a child at a table with a single marshmallow on a plate. The experimenter tells the child that he or she can either eat the marshmallow now or wait until the experimenter returns in order to receive two marshmallows. The unstated period of time to wait is 15 minutes during which the child is alone with the tempting treat. Those children who were able to exert self-control as four- or five-year-olds in the 1960s were found to have better academic achievement, monetary success, and even physical fitness several decades later. Mischel relates the cognitive strategy of self-control to the ability of the pre-frontal cortex in the brain to inhibit the urges of the more primitive limbic system. By generalizing to the Moral model of addiction, is the lack of sufficient will power simply a result of poor pre-frontal cortex functioning? There is ample evidence of the pre-frontal cortex role in governing impulse control and executive function in the brain. This leads to the next model for discussion.

In contrast, but not altogether contradictory to the Moral model, is the Disease model. In contemporary terms, this model focuses on neural dysfunction—bad brain chemistry or circuitry—due to the drug and, consequently, creates a dependency beyond individual control. Addiction specialists tend to adhere more to this view of addiction as disease. Note, however, how this model can be overlaid with the Moral model as indicating a weakness of the person: the lack of will power is due to a defective brain, i.e., a dysfunctional pre-frontal cortex.

Past research on the personality angle focused more on underlying commonalities of the addictive personality with better established biological trait factors. The first theory to get scientific traction with this approach was offered in the 1950s by Han Eysenck in his three-factor model of personality. Eysenck proposed that the addictive personality was due primarily to the interaction of the three traits that made up his factor model: extraversion, psychoticism, and neuroticism.[25] Early work on trait theories of addiction provided evidence that characteristics such as extraversion and sensation-seeking were found in greater frequencies among addicts. However, needless to say, not all extraverts are potential drug abusers.

The Physical Dependence model is an older notion and one that is less favored now. It, too, focuses on physiology by emphasizing the power of withdrawal symptoms, thus leading the addict to seek the drug to alleviate that highly aversive body state. Drug addicts do indeed engage in drug-seeking behavior to avoid the unpleasant effects of withdrawal. The withdrawal symptoms tend to be the opposite of the particular drug action. For example, the depressant action of raising the threshold of neural firing from drugs like alcohol, barbiturates and benzodiazepines can lead to seizures when a chronic user is suddenly taken off the drug. Opiate withdrawal is characterized by stomach flu-like symptoms with dramatic gastro-intestinal disturbances. However, the emphasis on physical dependence is not very helpful in understanding relapse and the long-term cravings of the chronic drug user. For example, the model is unable to account for cravings long after detoxification and withdrawal symptoms have subsided. Contemporary texts on substance use and abuse now use preferred language of dependence—physical, psychological, or both—in order to avoid some of the contradictions in the verbiage of addiction. As a singular theory to explain addiction, the Withdrawal model has fallen short.

Finally, the Positive Incentive model examines the role of neural and behavioral reinforcement: addictive drugs exert their influence via the brain's reward pathways and cause long-lasting changes in those pathways.[26] This is the model currently receiving most attention of neuroscientists due to advances in the neuro-pharmacological and brain imaging research regarding the dopamine pathways in the midbrain. Specifically, reinforcement pathways mediated by dopamine in the ventral tegmentum in the midbrain become altered with the use of certain drugs and in effect become hijacked. Without the drug, the reinforcement pathway no longer functions normally. Numerous studies now point to dopamine being the neurotransmitter central to the brain's reward pathway and relevant neural pathways have been found that connect the midbrain with other brain areas associated with addiction: the nucleus accumbens, the amygdala in the limbic system, and ultimately the frontal lobe of the cerebral cortex. Those areas are involved with reinforcement, emotion, and conscious thinking and planning. The Positive Incentive model more adequately explains the drive to take the drug and the cravings that can aggravate the user long after cessation of the drug.

Note that none of the models are mutually exclusive, although the Physical Dependence model is the most deficient in explaining the clinical experience of addiction. The Moral model similarly has been unhelpful both from a prevention and therapeutic standpoint yet remains popular in our vernacular and in conservative camps. Each model has its defenders and enjoys varying levels of support among researchers, addiction counselors, public policy makers, and the public at large. All attempt to assign blame to either the drug or the person. Is it truly a zero-sum phenomenon or could addiction more likely be an interaction?

Let us now turn to another period of time when addiction became front page news. An examination of addiction during the Victorian era can reveal some striking similarities in discourse and perhaps shed light on which paths are more likely to yield positive results in the quest of solving this public health problem. This was the time that opiates underwent a "Jekyll-to-Hyde metamorphosis" from household remedy to vilified scourge of society.[27]

The second half of the 19th century also saw epidemic levels of accidental deaths due to opiate overdoses. While the public health statistics are virtually impossible to compare directly then and now, one source puts the 1899 accidental overdose incidence somewhere between 0.3–0.5 percent of all accidental deaths.[28] That figure pales in comparison to current numbers where opiate-related deaths now account for the majority of accidental deaths, especially among the adolescent/young adult demographic. However, the problem was significant enough to catch the attention of the medical community, the public, and ultimately, the politicians of the day.

Opium, of course, had been used therapeutically for centuries since ancient times. It became a household item with the advent of laudanum, a preparation of opium in spiced wine, and enjoyed widespread use. Physicians and pharmacists dispensed it liberally and it could be obtained cheaply from any grocer or pharmacist in the form of various over-the-counter patent medicines. It was probably the most important standard remedy for illnesses ranging from headache to tuberculosis. As now, there was no more effective analgesic available and it was critical in the treatment of childhood diseases for which diarrhea would ordinarily result in dehydration and death. This is because one of the potent side effects of opiates is constipation. Given the widespread use among the general public, it was not unknown for some individuals to develop a habit of "opium eating." The term refers to opium in its pure form being sold in penny bricks; nonetheless, the majority of consumers, including the so-called category opium eaters, consumed opium in liquid form. The habit was considered a vice in the same way as tobacco use and not even considered as serious as inebriety caused by alcohol. Until the latter part of the century, there was simply no concept of "addiction" as we know it today and it was certainly not considered a medical issue, as evidenced by the lack of references in the medical journals. Yet within a few decades, there was a complete shift in public attitudes regarding both opiate drugs and their users.[29]

The debates surrounding opiate use were complicated because of the diverse dispensation and administration practices and the fact that there were very few alternative therapies for the ailments they purported to treat. Why would one knowingly restrict the use of a class of drugs that was ubiquitous and enjoyed strong public and professional support?

The initial medical discussions focused on the effect of opiates on longevity—an indirect nod to the overdose question. Prior to the 1880s, medical journal articles about opiates were more concerned with questions of toxicity and accidental poisonings. A famous insurance trial in England following the overdose death of the Earl of Mar in 1828 brought some of these issues to professional and public view. The case involved the payout of £3000 by the Edinburgh Life Assurance Company. The company had refused to pay the sum on the grounds that the Earl was a habitual user of laudanum and alcohol.[30] The key question before the court was whether the non-disclosure of an opium habit should invalidate the insurance claim. A related issue was whether an opium habit had known direct negative effects on longevity—there was nothing in the medical literature at the time to support that view. The Earl had exhibited no outwardly detrimental symptoms of his drug habit and Christison, a physician himself writing about the case at the time, states, "It does not necessarily follow that the habitual use of narcotics must tend to shorten life."[31] According to Foxcroft, the case was the first public debate on the moral issues concerning the vice of opium-eating, individual differences (e.g., personality), sociocultural elements that conflated with morality with social class standing, and the physiological consequences of the drug itself—in short, the same debates we have today concerning drug use.[32]

But the real shift in attitudes towards opiates can be pinpointed to several factors that emerged during the following decades of the 19th century: an increase in accidental infant overdoses, the differentiation of drugs and poisons, increased specialization of medical professionals, the popularization of hypodermic needles, addiction among physicians themselves, and the publication in 1878 of Levinstein's *Die Morphiumsucht.*[33]

As opiate drugs increased in home usage, there was a concomitant marketing of drugs for treatment infant illnesses and conditions like teething discomfort. Entrepreneurs and physicians alike made over-the-counter remedies laced with opium, as well as other drugs that are now considered narcotics. These patent medicines, so called because of their proprietary formulas were inexpensive and ubiquitous. There were no labeling laws for medicines and brand names gave no hint to their actual ingredients. Mrs. Winslow's soothing syrup, Dover's powder, and Godfrey's cordial were but a few of the many popular remedies in the Victorian age. These were considered unrivalled medications in the treatment of ailments like teething and diseases such as diphtheria, dysentery, measles, bronchitis, and tuberculosis.[34]

Early warnings appeared in medical journals as early as 1808, where it was posited that upwards of 25 percent of infant deaths were attributable to opiate abuse.[35] Public sentiment, however, was not stirred until infant deaths due to opiate overdoses began to rise in the middle decades of the century and the print mass media began to sensationalize it.[36] Opium was "an infant quietener," able to soothe all manner of ills and an effective sleep agent.[37] While the practice was widespread in both the United States and Britain, it was most notorious among working-class mothers who could ill afford doctors' fees but could manage the opiate-laced drops or syrups that were available for a penny or less. In some cases, the opiates were given to induce sleep, giving the exhausted parent(s) respite. The estimated doses sold for the sole purpose of doping children were staggering. One English pharmacist alone in 1871 was reputed to have sold six (imperial) pints a week for children when dosages for children were measured in teaspoon fractions.[38]

Morphine, a derivative of opium, was first isolated in 1805 and liberally added to the arsenal of opiate concoctions and patent medicines but it was the invention of the hypodermic needle in 1853 by Alexander Wood that enabled the drug to create its own addiction problems.[39] The use of the hypodermic needle for morphine administration in the 1850s meant higher doses of the drug were being delivered compared to the much smaller doses used for laudanum, much of which would be absorbed by the gut and metabolized before ever reaching the brain.[40] Initially, the medical profession encouraged patient self-administration with hypodermics. However, as physicians realized the potential loss of professional control combined with the marketing impact of patent medications, they began to lobby for exclusive powers over the method of injection. Morphine addiction then became more "therapeutically induced" among the middle and upper classes.[41]

Working-class individuals typically could not afford doctor fees and self-medicated with inexpensive over-the-counter opiates, such as laudanum. In contrast, individuals higher up the social class ladder could afford doctor consultations, as well as the higher cost of morphine and a syringe. Physicians routinely prescribed morphine for menstrual pains and thus gave women a socially acceptable yet private way using hypodermic needles to self-medicate while men could use opium publically in the dens and later dance halls and pool rooms that opened in cities when heroin came on the scene.[42] English physician Crothers notes in 1902 that the first warnings of morphine addiction were published in the medical journal *Lancet* in 1864.[43]

English-speaking physicians were then exposed about 15 years later to the English translation of Levinstein's *Die Morphiumsucht* or *A Morbid Craving for Morphine.*[44] This was the first scientific analysis of morphine addiction, a work that systematically laid out the disease state of addiction. Significantly, Levinstein defined morphine addiction as a disease.[45] Here was a book that contained chapters on symptoms, etiology, treatment, and prevention—just as any medical book focusing on a disease would do! Levinstein had further differentiated "morphinomania" as a constitutional disease, a type of insanity caused by morphine use versus "morphinism," which was morphine dependence (but not addiction) iatrogenic in nature—in other words, a result of therapeutic use and easily reversed.

The book was published at a time when new theories of nervous system action were being promoted and the author couched his analysis in the brain science of the day. The nervous system was thought to be a fixed energy source. Rather than the drug action being simply a function of weak character, the opiate acted as both stimulant and depressant and morbid craving came from a defect of the electrical energy forces of the nervous system. Whether the individual effect at any given time was stimulation or depression was dose dependent.[46] Thus the disease was of somatic origin. Prior to the 1880s, addiction (other than alcoholism) had not been included in the *Index Medicus*; the first appearance of addiction as a disease category is in the 1883 edition.[47]

Levinstein's suggested treatment was basically the "cold turkey" approach, total withdrawal of the drug from the patient, isolation and monitoring of the resulting physiological effects.[48] Curiously, at least by today's standards of detoxification, chloral hydrate (a strong sedative) and unlimited amounts of champagne or brandy were allowed during the detoxification process.[49] In modern times, chloral hydrate was supplanted by benzodiazepines, such as valium, to ease withdrawal symptoms and prevent seizures. However, administration of any alcohol would be suspect.

At the same time, the Moral model came into play with the new developments in genetic science: addiction was a character issue but now due to genetic predisposition. The distinction between disease and vice was blurred. A theory of degeneracy first proffered in 1857 viewed the debilitating effects (inebriety) of opium and alcohol similarly.[50] While some might simply develop a bad habit of dependence, those with a constitutional lack of will developed morbid addiction, indeed a type of moral insanity or lunacy.

The concept of will or volition becomes the link between the psychological and the physiological pathology. "The idea that addiction was a disease entity came into ascendancy riding on the coat-tails of the 'rise of science,' and the new scientific model of addiction enhanced, and served to entrench, a moral view of opiate users."[51] This notion led to addicts being committed to lunatic asylums. An 1872 government report in England indicated that 20 percent of the asylum inmates were morphine addicts. Opiates became the cause of insanity whereas just a few years earlier they were the cure and were argued as such by medical professionals.[52]

Earlier, during the 1870s, opiate addiction was enfolded into a broader socio-medical discussion of inebriety.[53] While opiates were not considered as dangerous or debilitating as alcohol, they were increasingly viewed as drugs with the potential of causing similar problems. Interestingly, the medical community began to recognize the role of individual differences in drug response, potential for dependence, and probability of "cure." Addiction was a disease "embodied in the character of the individual and manifested itself through the will, which in this instance presented itself as a physiological aspect.... It was the quality of the nervous tissue, its stability or instability, which was all important in the resistance offered to temptation."[54]

Complicating the professional debates were the sociocultural variables at play: the role of class, gender, and race. Females were more likely to succumb to morphinism because of their weaker nervous systems and males more likely to be morphomaniacs. Indeed, in the 1890s, the typical opiate user was a middle-aged female taking morphine to deal with menstrual cramps or menopause, respiratory illnesses, and general anxiety (hysteria).[55] Kandall indicates that by the end of the 19th century, two-thirds of the opiate addicts in the United States were female and this figure was partly responsible for the rise of the pharmaceutical industry.[56] The female nervous system was thought to be weaker, less resilient to stress, and was manifested by weak will.[57] For women and adolescent girls, addiction was a disease, a product of their physiology. For males, taking opiates was a matter of personal choice.[58] It was not until the 1920s when the profile of the typical opiate abuser became a young adult, working-class male.[59]

During the middle Victorian era, addiction was primarily a problem among the professional and upper-classes.[60] They were the individuals under stress, had access to physician-led medical care, and could afford morphine. The working classes were not part of the equation as they could only afford the cheaper laudanum and patent medicines and had little to no access to professional treatment. In his medical discourse, George Beard blended the moral model with the late Victorian era neuroscience: opiates drained the nervous energy in those working-class individuals who by virtue of their social class lacked moral fortitude.[61] Paradoxically, he treated the "opio-maniac" with more opiates. Despite a lack of evidence, there was also a notion that working-class individuals took opiates recreationally while middle- and upper-class individuals only took them therapeutically.[62] As greater industrialization, urbanization, and new immigration patterns emerged with the Progressive era in the early 20th century, so did the image of the opiate addict change. In the United States, this re-imaging took the form of vice, deviance and degeneracy.[63]

The race card came into play in relation to the Chinese immigrant population: dock workers in East London in England and in San Francisco in the United States. The Chinese immigrants established the infamous opium dens popular in both those cities. The opium use that was recreational, rather than therapeutic, subsequently came to be associated with the Chinese and both print media and fictional accounts depicted the lure of the exotic and foreign. In Britain, campaigners against the opium trade, such as the Society for the Suppression against the Opium Trade, disseminated propaganda against the Chinese community to suggest that domestic opium use was associated with an "alien minority."[64] This played into fears of white female slave trade and cultural stereotypes of women needing protection. As anti-immigration and anti–Chinese sentiment grew, so did public pressure to close the opium dens and remit opium to illegal status.

Developments in the medical profession continued play a role as well. Recognition that the addiction problem was largely physician-induced resulted in pressure to reduce the number of prescriptions. Auren states, "Addiction itself was not new—doctors' role in addiction was."[65] As the medical community continued its course towards increased specialization with new divisions between neurology and psychiatry, subspecialties of expertise in addiction also arose in the 1870s. These developments were exemplified by new physician-led societies in the inebriety movement, such as the American Association for the Cure of Inebriates in the United States and the Society for the Study and Cure of Inebriety in Great Britain.[66] First, as a result of their lobbying efforts, legislation was introduced to restrict the use of hypodermic needles to medical professionals. Then, as new analgesics and sedatives, such as aspirin and barbiturates, were developed in the waning years of the 19th century, physicians at last had alternatives to substitute for the opiates. It was this development more than any other factor that finally led to significant reductions in opiate addiction by the end of the Victorian era. Passage of the Poison Act in Britain and the Pure Food & Drug Act in the United States were the legislative outcomes of this public health campaign. Prescription opiate addiction declined after a peak in the 1890s, only to be replaced by dependence associated with recreational and illegal use.[67] Finally, the use of opium and its derivatives was relegated entirely to the category of vice and deviance and those drugs were erased of their initial therapeutic veneer.

Let us look at the parallels between our two eras. In both periods, prescription opiate addiction began iatrogenically—a result of therapeutic administration. Only later was there a shift to drug usage primarily for recreational purposes and, as legislative or professional controls on prescription medication tightened, to now illicit drugs, such as heroin. During both periods, individuals either self-medicated or were prescribed various opiates mostly for pain management but certainly a wide variety of health ailments in Victorian times. In some cases—but not all—that practice gradually resulted in addiction. Addiction tended to develop in situations where dosages were increased, such as through hypodermic administration. In both periods, government officials were slow to heed the warnings coming from the medical and public health sectors and when they did react, did so by tightening controls on access to the narcotics indiscriminately. The stricter regulations, however, had the unintended consequence of increased crime, emerging black markets, and suffering by patients truly in need of the pain relief only attainable by opiates. Overreaction to one problem created new and unforeseen drug problems. Then, as now, cheap heroin (and now fentanyl) replaced the opiates that had been once therapeutically administered.

There were gender differences with respect to addiction and overdoses in both periods, but not in the same way. In Victorian times, more males than females were considered addicted; opiate addiction in females was a disease of the nervous system. In contemporary times, females are overdosing with prescription pain killers at greater rates than males. However, conflation of moral and disease models and differential application by gender probably mitigates real gender differences in addiction.

The disease models of each era offered explanations that were consistent with the physiological narratives of the day and both periods still had their respective versions of the moral model. The contemporary version is evident in the widespread but ineffective drug prevention programs Just Say No and DARE. Resisting temptation is the cornerstone of conservative approaches to addiction. Unfortunately, empirical evidence does not give support to the effectiveness of the moral model.[68] The Victorians did have a peculiar spin on their theories of addiction with applying the moral model to males, while the disease model explained female addiction. This allowed them to keep gender roles entrenched within traditional paradigms.

Finally, the seminal question about whether opiate addiction is a function of psychology or physiology has been debated for 150 years. The very concept of addiction in relation to opiates and other psychoactive drugs is not simply a physiological entity or based on empirical science but one socially constructed and often applied in response to societal fears about a particular minority ethnic group or social class.[69] Public discourse for both camps has been eerily similar yet ultimately led to different approaches in the United States and the United Kingdom. Britain's public policy has traditionally relied on the disease model, coming out of the Victorian era. Addiction is considered a medical condition requiring prevention, treatment, and in some cases, maintenance. In contrast, American public policy has traditionally used the moral model of addiction and thus criminalized the behavior. Research coming out of contemporary neuroscience regarding the reward pathways in the brain may effect a shift in policy to take a more "brain-centered" approach. Commonsense practices like equipping first responders with naloxone, the drug that reverses opiate action within minutes of administration, will reduce the number of deaths by opiate overdose.[70]

How will this play out with addressing the current prescription opiate crisis? Time will tell but what the history of opiate addiction teaches us is that the tendency to view addiction through a dualistic lens separating mind and body inevitably means that important variables in a complex process are ignored. In both instances, there has been a reluctance to incorporate dynamic, interactive models that include behavior, personality, and physiology that operate in a sociocultural context. The actual pharmacological and medical science related to these drugs are often given a back seat when it comes to public policy and legislation. When a holistic, monist approach is tried, perhaps real gains in fighting addiction will be achieved and policies on restricting opiates for therapeutic use will be based on science and not politics.

8. You Are What You Eat: Gut and Brain

Hosting a dinner party in today's society is fraught with dietary challenges. Guests may be vegetarian, vegan, pescatarian, and/or follow gluten-free, paleo, or any number of popular diets. Individuals may have life-threatening nut or seafood allergies, celiac disease, or be lactose-intolerant, all of which results in gastric discomfort. Some might argue that these various dietary fads are an outgrowth of improved standards of living and vast technological improvements in agriculture. In other words, dietary choices are both limited and enhanced by the variety available in our food environment, socio-cultural norms, and individual taste preferences and philosophies. Some of the aforementioned diets are considered quite mainstream now and supported by the standardized grocery chains and restaurants, while others may still be viewed askance by some.

But you might be surprised to learn that even the most common today of these dietary variations, vegetarianism, was also once viewed not only as unnatural but also with suspicion by the medical establishment perhaps because it was often in conjunction with other questionable fringe practices, such as spiritualism and mesmerism.[1] Its adherents tended to be zealots and promoted the practice as a "panacea for societal and moral ills."[2] Followers of Grahamism, an early 19th-century American health movement inspired by Presbyterian minister Sylvester Graham (whom graham crackers were named after), believed that all disease was linked to poor diet and particularly the consumption of refined flour. White flour in the 19th century was commonly cut with additives, such as alum or chalk, to bulk it up, so there was more than a kernel of truth to the health problems associated with white bread.

Graham advocated a vegetarian diet, as well as temperance, for healthy living. That notion is not controversial at all today perhaps but then Graham's health concerns did not take the same form as contemporary nutritionists. For example, Graham was especially concerned with the problem of adolescent masturbation (and its outcomes of insanity and life-long depravity).[3] Vegetarianism was a preventative measure against the practice. Gregory notes that utopian movements in the late 18th and early 19th centuries often had a dietary component to them.[4] The overlap of dietary fads and rejection of medical orthodoxy led 19th-century physicians to lump vegetarianism, even if only a "harmless delusion," with other quackery.[5]

A View of the Nervous Temperament, published by Thomas Trotter in 1807, laid out a litany of behaviors of modern living that was contributing to the near-epidemic proportion of nervous diseases.[6] Diet was prominently featured, much as it had almost a century before in *The English Malady* by George Cheyne.[7] In fact, at that time Cheyne had espoused a vegetarian diet for good health; it turns out that this was an easy proscription to make at a time when only the wealthy could typically afford meat on any regular basis. For Trotter, however, the benefits of healthy living were couched in what would become a hallmark of the Romantic era: a return to nature ("back to the land") and simpler times.[8] Readers were admonished for their rich diets and encouraged to eat more fresh fruit, meat (so not a vegetarian diet here) and dairy and to avoid the spicy foods that were now appearing in the English diet, thanks to trade with the West and East Indies. Above all, individuals (especially females) should avoid drinking tea, which could lead to nervous debility.

Vegetarianism was often associated in the Victorian era concurrently with the alcohol temperance movement.[9] Both social movements were concerned with promoting "self-help, self-control and social progress through individual reform."[10] Vegetarianism was temperance taken to an extreme as it represented a departure from the social norm that went beyond the prohibition of ingestion of just one substance. "Why should temperance be limited to drink," Sylvester Graham wondered.[11]

The post–modern era has seen the same nutritional myths that all disease results from improper diet: everything from poor eating habits to additives and toxins in over-processed foods have contributed to ill health.[12] Only the consumption of "natural" foods leads to healthy living, according to these proponents. The goal for health, often straddling the line between science and pseudoscience was "salvation through nutrition."[13] The deregulation efforts in the 1970s and '80s actually undermined the original intent of the Pure Food & Drug Act, making it easier to add "non-natural" ingredients to the food supply. This has resulted nowadays in a backlash against food additives and a proliferation of organic and "natural" (which actually means very little since it is not a clearly defined or regulated term) products. That deregulation of nutritional guidelines at the same time removed vitamins and nutritional supplements from the Food & Drug Administration's regulatory purview.[14] Now there are countless substances, often unlabeled and certainly unregulated for quality, that are promoted as detoxifying agents. A recent internet search using the phrase "brain detox supplements" yielded over 18 million results.

The purity and cache of vegetable diets spread to ideas about brain health and illness even in the first half of the 19th century but the manifestation of these ideas did not always engender much success. In the mid–1830s, the medical communities in the United States and Great Britain began to take notice of a rise in deaths among patients who had dosed themselves with a particular patent medicine, Morison's Pills. Report after report was published in the leading medical journals of both countries. *The Lancet*, one of the foremost British medical journals then and now, contained six different reports of fatalities or court cases against the maker in the 1836 volume alone, as well as three editorials on the subject. The fatalities all occurred in patients who considered themselves followers of "Morisonianism"—devotees of the vegetable pills manufactured by James Morison, president of the British College of Health.

Morison's pills, an early Victorian detox agent, were advertised widely as a panacea, a cure-all for all ills, including nerve and brain maladies. The pills could even cure mental derangement! The pill, actually named "The Hygeian Vegetable Universal Medicine," was basically a concoction of various purgatives to rid the body of toxins and cleanse the blood. The success of the so-called vegetable medicine was predicated on the greater purity and efficacy that a vegetarian diet supposedly held.

James Morison (1770–1840) was no physician but rather a Scottish businessman, who created an extremely successful franchise that tapped into the public's health fears. His personal motivation purportedly stemmed from a previous life of ill health with constant suffering that received no relief from traditional medical therapies. According to his own report, by age 50, he had consulted over 50 doctors and tried remedies that included a variety of pharmaceuticals, exercise, hydrotherapy, leeches and bloodletting, and even steel trusses and surgery to alleviate tightness in his chest.[15] His symptoms were consistent with what would later in the 1870s be considered as neurasthenia (nervous exhaustion) or hysteria.

Retired (or perhaps invalided out) from a successful career in the wine and spirits trade of the West Indies, Morison decided the cause of his symptoms was that his bodily humors were out of balance. Bile from his stomach and digestive tract were diffused throughout and his blood was filled with toxins. He was further distressed to have been treated with chemical agents, such as mercury and ether. He then reasoned that some sort of vegetable cure was needed to detoxify his blood and thus developed his patent formula and "The Hygeian System."[16] Sale of the pills was launched in 1825 and advertised as consisting of pure vegetable matter in contrast to the toxic chemicals prescribed by traditional chemists (pharmacists), apothecaries, and physicians.[17] Subsequent chemical analysis of Morison's Pills revealed that the formulary was in fact vegetable matter but it was an admixture of mostly natural laxatives: aloes, jalap, gamboge, cococynth, and rhubarb. The idea was "purgation by vegetables" to cleanse the blood.[18] Over the years the formulary changed slightly, at times including other ingredients like senna, cream of tartar, and myrrh. The exact ingredients or proportions were never divulged officially despite numerous requests by the medical establishment so the remedy could be subjected to appropriate testing. Morison defended his secrecy by saying that he feared imitators getting hold of his successful formulation.

The pills came in two forms, No. 1 and No. 2, with the latter being of greater strength than the former. The No. 1 pill was billed as a "mild aperient" and No. 2 pill was the "purgative."[19] The typical recommended dose was 15–20 pills a day; if that dose failed to provide relief, taking even more was urged by Morison.[20] The dosages varied according to the condition being treated but not in any discernable logical fashion to the medical observer. Doses might alternate between the two forms of the pills and had any number of permutations of order, number, and timing. They were indeed strong cathartic purgatives and Helfand notes their similarity to Anderson's Scots pills, which had been on the market since the 1600s![21] The "macro" dosing was a deliberate and distinctive feature of the Hygeian method. If the dose were too small, it would not be sufficient to detoxify the blood. Patients were to continue taking the pills even after symptoms began to subside to ensure that "all corrupt humours in the body were properly expelled."[22] Similarly, if the symptoms failed to abate, that simply meant that more pills were necessary to affect a cure!

These pills were purchased from "agents" rather than chemists or pharmacists and the pills were sold for more than a hundred years in several countries. The British College of Health, located on a main thoroughfare in London, was not an educational institution as the name might suggest but rather a building from which Morison ran his very successful business. From here he published numerous tracts and articles promoting his Universal Vegetable cure and railed against the medical establishment. His proposed system of health, the Hygeian method, was premised on the following 10 points:

> (1) The vital principle is contained in the blood. (2) Blood makes blood. (3) Everything in the body is derived from the blood. (4) All constitutions are radically the same. (5) All diseases arise from the impurity of the blood, or in other words, from acrimonious humours lodged in the body. (6) This humour which degenerates the blood has three sources, the maternine, the contagious and the personal. (7) Pain and disease have the same origin; and may therefore be considered synonymous terms. (8) Purgation by vegetables is the only effectual mode of eradicating disease. (9) The stomach and bowels cannot be purged too much. (10) For the intimate connection subsisting between the mind and the body, the health of the one must conduce to the serenity of the other.[23]

So Morison's Hygeian System was clearly coming theoretically from the humoral school of medicine and espoused a "single cause" for a "single problem": the problem was bad blood.[24] Morison particularly eschewed the more modern (i.e., Enlightenment era) medical view of "sympathy" between afflicted organs and, as indicated in his most popular 1828 publication *Origin of Life*, denied the brain's role in nervous diseases.[25] While merely one of innumerable quacks touting fringe medicine in the early Victorian era, Morison stood out because he was selling not just a nostrum but a belief system. Indeed, he was referred to as the "King of the Quacks" by Dr. Thomas Wakley, editor of *The Lancet*, and especially singled out for personal admonition.[26] His was a system of holistic health and like many other alternative medicinal approaches that emerged during this time period, it filled a public need that was unsatisfied as traditional medicine was further embracing evidence-based science. Georgian medicine had been very much based on the interpersonal interaction with the physician and, according to Porter, more patient-centered.[27] As medicine became more expert-centered at the wane of the Georgian era, several alternative medicine movements gained popularity, some of which continue even today. These movements included homeopathy, naturopathy, hydropathy, phrenology, and mesmerism. "Each of these movements proclaimed itself to be in possession of a more subtle understanding of the true nature of disease as an integral part of the active processes of Nature ... and claimed to use more natural modes of healing."[28] However, Morison's approach followed the methods that were the hallmark of quackery and patent medications: secrecy in regard to ingredients and formulation, discrediting of the competition, contempt for traditional medicine, and heavy use of personal testimonies.[29]

Morison's system was also distinctive, however, by his marketing approach and the use of what is essentially a franchise system.[30] He mimicked traditional medicine and the Victorian love of learned societies with his "college" of health that oozed respectability. He would train and license "agents," who then peddled the pills. For years, the pills were available for purchase only through the College or these agents, who were salesmen, not pharmacists; only in the 1850s did the pills reach the public via the shelves of pharmacists and druggists. Morison gave public lectures where free samples of his pills would be distributed to the audience. He self-published a stream of illustrated fliers, health tracts, and pseudo-medical journals that were filled with screeds against traditional medicine, endless testimonials, satirical cartoons aimed at the failings of physicians and modern medicine, and loads of advertisements. *The Hygeian Journal* (later *The Hygeist*) and *The Anti-Lancet* were the most prominent of these self-made publications but Morison also took out advertisements in existing popular magazines and newspapers. Helfand notes that even respectable newspaper editors who had earlier criticized Morison's pills, changed their tune once Morison boosted their revenues by buying substantial amounts of advertising space.[31] Another publicity tactic was to introduce petitions against poisons and medicine into the British Parliament. One petition in 1847 had 20,000 signatures![32] Within 15 years, the pills were sold in at least six countries and Morison's income was reputed to be a phenomenal £80,000 (approximately $110,000).

The series of lawsuits filed against Morison's pills and counter-suits by Morison in the mid–1830s in addition to the increasing reports by physicians of patients who had died while taking the supposed medicine yielded negative publicity for the patent medication. A report in the *Boston and Surgical Journal* in 1836 indicated, "Scarcely a day passes without instances occurring of a serious mischief from the preposterous use or abuse of this quack medicine!"[33] Several of the medical case studies published in the *Lancet* painted a picture of loyal patients who took more and more pills at the behest of their Morison's agent despite not only failure to improve but in the face of obvious physical decline and absolute prohibition by the patient's physician! While a laxative agent could indeed relieve suffering of some ailments, these strong purgatives further weakened the patient's constitution to fatal levels of dehydration. Some physicians noted that the fatalities were not caused by a particular toxin found in the pills but the sheer quantity and indiscriminate use of them.[34]

Reports of patients taking hundreds and even thousands of Morison's pills in a year were commonplace.[35] When patients got worse, Morison and his agents prescribed even more pills and discounted the worsening symptoms as just being signs that the disease was being "drawn downwards from the head."[36] As Morison's vice-president, Thomas Moat, says, "If two or three pills do good, five or six do more."[37] In the courts, however, it was the agent who was found to be negligent and not the product or James Morison himself.[38] Dr. Wakley wondered why was Morison not on trial if pills prepared under his direction and supplied by him to the vendor. It was an injustice to punish the agent while the "principal go scot-free."[39] He also noted that had Morison or his agent been an apothecary, they would have been punished by the court for operating without a license. Yet the laws were such that Morison could operate with impunity.

The medical outcry against Morison's enterprise was part of a larger campaign against quackery, of which Thomas Wakley was a leader. In reading the various editorials and letters to the editor by physicians, one can readily perceive the frustration as the medical community tries to exert its authority against these populist trends. Wakley (1836d) notes that it is sometimes impossible to differentiate between "irregular and regular" medicine but a start is to look at the extremes.[40] Morisonianism was illustrative of the irregular extreme in the continuum. Other forms of quackery, such as homeopathy, were more difficult to identify. How should the medical profession or courts proceed when even their queen (Queen Adelaide, spouse of King William IV of Great Britain) was a devotee of homeopathy? Would the courts actually send her German homeopathist to prison?

Morison decamped to Paris in 1834 after his court cases were going badly for him but the business operation continued for almost century afterwards.[41] A statue on the grounds of the British College of Health building was erected from penny subscriptions and remained until torn down when the road was widened in the 1920s. Morison's pills were integrated in popular culture, featured in songs, satires, and caricatures. Simple entrepreneurship had shifted from simply dispensing pills to selling an ideology.[42] While the medical establishment had some successes in curtailing use of patent medicines like these and spearheaded regulations that would eventually lead to the passage of the 1906 Pure Food and Drug Act at the turn of the 20th centuries, the success of Morison's pills had been "compounded by the admixture of chronic hysteria and masochism" in society.[43]

Morison's pills were touted as a natural remedy because the ingredients consisted wholly of vegetable matter. As it turns out, this was a precursor of the broader vegetarianism movement and other vegetable remedies in the patent medicine trade. Some of the features of the Hygeian system were hardly new, such as grounding the theory in the ancient Greek humoral framework. Likewise, even though not advertised as the active mechanism in the Morison's Pills as such, the notion of purgatives to cleanse the body and spirit was a long-standing medical tradition, going back to the writings of ancient Greek physicians. Galen, for example, wrote about the relationship of mental disturbance and symptoms suggesting dysfunction of the gastro-intestinal tract.[44] The saying "All diseases begin in the gut" is attributed to Hippocrates.

Even though the medical establishment railed against the quack vegetable patent medicines, the field was itself still deploying emetics and purgatives from their therapeutic drug arsenal. As the asylum movement in the mid–19th century gained popularity, American alienists and physicians amassed data from autopsies of mental patients.[45] Notable in these data were observations about irregularities in the gut, often in the absence of visible lesions in the brain. Some physicians began to propose a link between the two, just as the ancient Greeks had done. In a paper read in 1863 before the American Association of Superintendents of American Institutions for the Insane, Workman noted that no one who regularly worked with the insane could fail to notice the association of gastrointestinal dysfunction and insanity. Furthermore, the condition was usually made worse by overuse of "drastic purgatives" and cathartics. He strongly believed that restoration of normal bowel function was a first and necessary step in the cure of mental afflictions.[46] However, despite the purported relationship between the gastrointestinal system and mental illness, these physicians did not specify a particular diet either as prophylactic or treatment.

In contrast, some 50 years later during the "Age of Anxiety" in the late 1800s when neurasthenia was a leading diagnosis, neurologists weighed in on the relationship of diet and nervous disorders. George Beard, the physician responsible for laying out a systematic theory underlying neurasthenia, believed that a vegetarian diet was not only foolish in general but downright deleterious for the nervous system.[47] His rationale was based on the new evolutionary theory that was finally permeating science a few decades after Darwin's seminal publications. According to Beard, humans should primarily eat products (meat and dairy) from animals that were closer to humans on the phylogenetic scale. This would ensure ease of food assimilation by the digestive system. Grains (cereals) and fruits that are the furthest removed on the phylogenetic scale likewise should be avoided. The best foods for adjunctive treatment in nervous diseases were "beef, mutton, lamb, chicken, eggs, milk, fish, butter, and wheaten bread."[48] No fruits or vegetables should be consumed nor fats except butter, beer, ale, and most wines (the beverages due to their sugar content). Decreased consumption of "stimulants" were also advised: less coffee, tea, brandy, and claret. Only "muscle workers" (i.e., laborers) should eat pork.

Silas Weir Mitchell, famed American neurologist and author of the Rest Cure described in Chapter 12, had a particular interest in the dietary issues related to nervous exhaustion. In his 1898 best-selling work, *Fat and Blood, and How to Make Them,* he directed most of the discussion to the clinical cases of women diagnosed with nervous exhaustion, "spinal irritations," and hysteria. Besides the various psychological and somatic symptoms presented in these cases, the women were usually underweight and, using today's term, anorexic. Mitchell's approach, however, is distinctive among his peers in that he is not recommending diet as merely an adjunct to the "real" therapy nor does he rely on diet alone to treat the nervous ailments. For Mitchell, it is part of a holistic theory that marries body and mind.

The basic elements of Mitchell's Rest Cure included seclusion, rest, massage, electrical stimulation of muscles (to prevent muscle cell atrophy caused by the rest component), and a specific diet to encourage the development of fat to what was considered a healthy level and improve the blood from its anemic state.[49] Another critical feature was the detoxification of the body and brain from the host of previous pharmaceutical treatments, either prescribed or by self-medication. Like so many physicians in the decades, even centuries, before him, Mitchell noted the high comorbidity in nervous complaints with gastro-intestinal upset, such as dyspepsia. This was the observation that led him to make milk consumption the cornerstone of the rest cure diet.

The milk consumed ideally should be skimmed of its fat and as fresh as possible—in fact, straight from the cow, twice a day was best![50] However, Mitchell recognized that not all patients would have this kind of access to fresh milk, so pasteurized milk that could then be warmed before drinking was an acceptable substitute. He prescribed a rigid regimen of administration of the milk, beginning with four ounces every two hours and gradually increasing both the quantity to two quarts and the interval to three hours. At the beginning, slight amounts of salt, caramel, tea, or coffee could be added for those patients who found the taste of pure milk intolerable. This exclusive diet was typically employed for the first week or more of treatment and while initially there would be weight loss due to the small quantities and lack of exercise, gradually the trend would reverse and patient would begin to gain weight.

In the case of the neurasthenic, hysterical patient, the milk diet itself is responsible for the detoxification of what was then a typical drug regimen of morphine, chloral hydrate, bromides, and more.[51] Mitchell believed that in many instances, the combination of medicines was simply masking symptoms and creating its own set of problematic side effects. While milk was the central ingredient of the diet (and hence the name "milk diet"), solid foods and malt extract were gradually introduced. One unusual addition was a cold beef broth: a pound of raw beef in a pint of water first chilled overnight, then kept at a tepid temperature for two hours, and subsequently strained. The resulting "soup" was an uncooked, raw beef filtrate. Fats allowed were copious amounts of butter and then daily doses of cod liver oil. Iron supplements to correct the anemia of the blood were also added to the diet. As the treatment progressed, more and more foods were introduced and unlike those prescribed by his theoretical predecessor George Beard, Mitchell did not prohibit fruits and vegetables. The milk diet expressly targeted gastro-intestinal dysfunction exhibited by Mitchell's patients but he did not perceive, or at least comment, that the dysfunction might be directly related to the nervous condition.

Only a few decades later, pediatrician Sydney Haas published a paper with his clinical findings that a carbohydrate-free diet with the sole exception of bananas could cure celiac disease.[52] Despite his research consisting of only ten clinical case studies in which treatment occurred within an uncontrolled environment, his "banana" diet caught on quickly. Abel in 2010 notes that Haas' research fit nicely into food fads that paralleled medical quackery seen during the patent medicine era.[53] Levenstein (1988) refers to the decades bracketing the 1900s as "the Golden Age of Food Faddism."[54] There were glowing personal testimonials, a singular remedy (e.g., the banana), and a miraculous cure that all the other practitioners of medicine using traditional means had somehow failed to discover. Haas' ideas, notably influenced by promotional research from the United Fruit Company, had a scientific cast to them and he continued to document success with his treatment. In 1950, his son and he published a study of more than 600 children, the majority of whom were successfully treated using his method.[55] They proposed that a new definition of celiac disease be adopted: "the successful recovery of patients following the adoption of the banana diet."[56] It was only a year later that the true cause of celiac disease was definitely revealed to be gluten and eventually a diagnostic test was developed to determine the presence of the autoimmune disorder. However, within a few decades, the diagnosis of celiac disease had all but disappeared until its revival in the 2000s.[57]

How is the Banana Diet relevant to this discussion of brain and gut?

Haas was primarily interested in its relationship with celiac disease. However, decades later a new fad diet would reference the Banana diet as a "specific carbohydrate diet" for individuals with a variety of neurological disorders, including autism. The 1960s brought with it a revival of food fads akin to the ones seen at the turn of the century. This "second Golden Age of Food Quackery" was precipitated by many of the same fears: concerns with artificial ingredients, additives, and mass production.[58] Any number of fad diets emerged in the following years but one diet in particular focused on a link between the nervous system and the gastro-intestinal tract and cited Haas' original study as inspiration.

The Gut and Physiology Syndrome, or GAPS, diet is purportedly a treatment for autism, ADHD, schizophrenia, and a host of other neurological or psychiatric disorders.[59] At the website by a key promoter of the diet, www.gapsdiet.com, the author makes claims eerily similar to those of James Morison of Hygeian System fame. Disease begins in the gut with food toxins contaminating the blood. The website notes that diverse neurological and psychiatric conditions, including "autistic spectrum disorders, attention deficit hyperactivity disorder (ADHD/ADD), schizophrenia, dyslexia, dyspraxia, depression, obsessive-compulsive disorder, [and] bipolar disorder," can all be treated by a detoxification of the blood. The website is replete with treatises and videos, almost all by the same author, and there are books and numerous products related to the dietary program available for purchase. In contrast to Morison's pills, however, the dietary emphasis is not vegetarian.[60]

The appeal of the pseudoscientific approach is bolstered by legitimate scientific research on the brain-gut axis, as it is currently termed. This connection between the brain and the gut has been a growing area of focus in neuroscience in the last decade. Once thought to be separate physiological systems, the gastro-intestinal tract and the nervous system have been found to have intimate communications between them. According to Rea, Dinan and Cryan, the axis is a complex network involving the central nervous system, autonomic nervous system, and the enteric nervous system.[61] Both neuroendocrine and neuroimmune elements interact among these various systems as well. At the core of the network is the gut microbiome, the world of microorganisms in our gastro-intestinal system. The collection of microorganisms—bacteria, yeasts, viruses and even parasites—make up the microbiota. The composition of the bacteria in our gut is largely shaped early in fetal development and happens concurrently with development of the nervous system.[62]

Environmental factors, ranging from vaginal delivery to exposure to stress can influence the development of the microbiome in early life.[63] However, it is important to note that the axis reflects bidirectional influence.[64] Information gathered from endocrine and immune cells in the gut affect the brain and the brain influences gut reactions. It may not sound far-fetched to posit a role of stress influencing irritable bowel disease, especially knowing how the autonomic nervous system can act on the gut via the vagus nerve.[65] But researchers today are actively examining the workings of the gut-brain axis in neurological and psychiatric disorders, such as schizophrenia, autism, depression, and even anxiety.[66]

As discussed previously, the relationship of gut and brain had been long observed by physicians who dealt with mental disorders from the era of the ancient Greeks to the 19th century. What was missing from those observations was an explanation of how seemingly disparate and remote physiological systems could even communicate. The key to understanding the relationship seems to be the neuroimmune and neuroendocrine systems. Inflammation that compromises the integrity of the endothelial and epithelial barrier in the gut wall precipitates bacteria translocation. This in turn causes the immune system to mobilize in defense, creating a vicious cycle.[67] The initial inflammation can be triggered variously by environmental factors, stress, toxins, genes, and infections. For example, exposure to the gut pathogen and parasite, *Toxoplasma gondii*, has long been associated with the development of schizophrenia. The infectious agent precipitates a response of food-based antigens and alters the but microbiome.[68]

Research studies have similarly looked at the relationship of the gut biome and stress. The stress response is well known as an important contributing factor in physical and mental health, as well as having generational effects, and similarly well known since the 1980s is the link between the stress response and the immune system. One recent proposition is that disruptions in the gut biome affected by the stress response can result in the transgenerational inheritance of that biome and thus influence the heritability of certain mental disorders.[69] A hypothesized model focuses on this interaction of gut, stress response (perhaps prenatal), and mental health. Interestingly, dietary interventions that affect the gut-brain axis have demonstrated generational effects with phenotypical expression of disorders, such as anxiety, in rats. It remains to be seen if the same results would be seen in humans with anxiety. Nonetheless, a corroborating picture of the role of the gut in brain function has emerged.

While there is still much to be done in elucidating exact mechanisms between gut dysfunction and brain disorders, this line of research may yield break-through approaches in treating multi-faceted conditions that perhaps reflect a conglomerate of developmental disorders, such as autism and schizophrenia. From genetic intervention, manipulations of food-related antigens to probiotic therapies, there may be yet be effective treatments for subsets of mental disorders that may actually have a dietary basis. This relationship of brain and gut has been written about for centuries, been subject to notions of quackery and various food fads, but the essentials of the relationship have shown to have some substance after all. However, instead of bogus vegetable pills, elimination of meat products, or the restriction of particular food groups, the magic ingredient to health may be the lowly bacteria and other microorganisms that inhabit our gastro-intestinal tract. The gut microbiome may hold the key to understanding that relationship and the essence of our being. Indeed, we are what we eat.

9. "The Heinous Sin": Viagra

BEFORE I BEGIN, let me caution the reader. This chapter not only deals with topics intended for a mature audience but will also take us down a path to mental and moral insanity, quackery, and ultimately, to a solution in search of a problem. The juxtaposition of current trends and theories both in neuroscience and psychopharmacology with their historical counterparts is a fascinating area for exploration and provides unexpected avenues that inevitably lead to new questions. For this particular chapter, I set out to examine the contemporary claim that the best-selling drug, Viagra was the first drug used to treat erectile dysfunction (or, as we all know from commercials, ED). If one sorts through the semantic nonsense, even a casual reader of that claim would realize that aphrodisiacs have been around for centuries. Whether truly efficacious or simply a placebo effect, the actual mechanism by which aphrodisiacs worked were largely undiscovered. However, what, if any, constituted mainstream drug therapy by the medical establishment and how did drug action fit within theoretical constructs of psychology or neurology of the day? The answers to this question took me in four different, seemingly convoluted, yet connected directions, which I will cover in the following order: (1) the story of Viagra; (2) impotence; (3) masturbation; and (4) sexual neurasthenia.

So first, let us recount the story or, should I say, pharmacological legend of Viagra. It is an interesting tale in its own right and one that has been examined before.[1] First marketed in the 1990s, the drug heralded a new age of diagnostic expansion, sexual dysfunction and "male enhancement."[2] It was one of the new "lifestyle" drugs which sought to enhance rather than treat bodily dysfunction. The actual drug, sildenafil citrate, inhibits the action of a particular enzyme, phosphodiesterase type 5 (PDE5), which normally acts to inhibit the chemical process necessary for the blood flow to achieve an erection. In other words, Viagra works by increasing and maintaining blood flow to the penis.[3] It is a mechanical action that bypasses the central nervous system (e.g., brain and spinal cord) and relies on physical forces of blood flow pressure mediated by the peripheral nervous system, the parasympathetic nervous system, to be exact.

Originally, sildenafil had been used in clinical trials by the Pfizer pharmaceutical company from 1989 to 1994 for treatment of angina because of its vasodilating properties. The company researchers at that time duly noted a side effect of increased erections. There was no intentional examination of the nitrate in relation to impotence at that time. The data from those angina studies were shelved by Pfizer until one of its pharmacologists in 1998 was mining old research studies for possible novel applications. Sildenafil was rediscovered; here was a drug in search of a disease. The Viagra story is, in fact, a tale of creative serendipity, not intentional development. The treatment not only preceded the disease, it preceded the science behind it![4] The notations regarding what was considered a side effect were suddenly interpreted in a new light. Figure and ground had shifted in this pharmacological perceptual illusion so that the side effect was now treated as if it were the main therapeutic goal.

The disease created for sildenafil was erectile dysfunction, a new label for impotence. Pharmaceutical therapies for impotence in the 20th century were indeed scarce, so Pfizer is not entirely misleading in its claims that Viagra was revolutionary among the various treatment options. Arguably, the drug did more for discovery of the physiological basis of erections than actually treating the cause or even understanding the etiology of impotence. There was relatively little understanding of the physics of an erection prior to that time and even less understanding for the causes of impotence. The two are linked, of course, but not conclusively so. After the advent of Viagra, researchers were able to show that a normal erection is a function of nitric oxide (NO) causing cyclic GMP (a biochemical second messenger) to relax the involuntary or smooth muscles of the penis via the parasympathetic nervous system. This allows blood flow in the area to increase and thus the erection occurs.[5]

At the time of sildenafil's rebirth, the correlation between degree of impotence and aging had been well established by the medical community. Despite the evidence published by paid Pfizer consultants that the majority of men over 40 years of age experienced mild impotence, this condition was deemed not "normal."[6] Was it really an abnormal condition in need of a treatment? The National Health and Social Life Survey of 1994 provided the original data on sexual dysfunction, attributing most of the underlying problems of impotence to psychological and interpersonal origins. These data were then re-analyzed by Pfizer consultants to arrive at the now standard frequencies of dysfunction in 43 percent females and 31 percent males. Prior to this, erectile dysfunction was notably a diagnosis in the *Diagnostic and Statistical Manual* of the American Psychiatric Association from its first edition in the 1960s through its fourth in 1994, indicating its status as a mental, rather than physiological dysfunction. Indeed, the term "erectile dysfunction" was originally a futile attempt to label impotence more firmly within the psychiatric camp—instead the label achieved just the opposite; instead of de-medicalizing the condition, it connoted a mechanistic problem of anatomy and physiology.[7] The National Institutes of Health sponsored a conference of primarily physicians in which consensus was achieved on the new use of the term erectile dysfunction. Urologists began to usurp the provenance of psychology with respect to treatment of male sexual dysfunction. With the advent of Viagra, Pfizer effected a major shift in attitudes and even medical explanations to create a category of "Sexual Medicine" and "Male Enhancement" drugs.[8] The drug not only preceded the science, it *drove* the science!

The interesting part of the story from a history of neuroscience perspective is how the underlying explanations shifted—not just the shift of psychological to physiological explanation in the 20th century, but a shift in etiological theory that occurred in the 19th century. For this we need go back a bit further to examine the proposed causes of impotence.

Prior to the Enlightenment age, the cause of impotence was firmly in the realm of religion: an organic sign of a deity's displeasure; medical references to impotence were rare.[9] Lack of fertility, as exemplified by impotence, was a "divine curse."[10] However, publications in the early 1700s ultimately influenced both the Church and the medical experts to reconceptualize impotence and its cause.[11]

In 1716, an anonymous manuscript was published that outlined the physical consequences of "The Heinous Sin of Self-Pollution." That sin was masturbation and here was a physical causal link with impotence rather than divine intervention. The term "onanism" stemmed from the reference to the sin of Onan in the *Bible*. The manuscript, *Onania*, was borrowed almost in its entirety from an earlier treatise of venereal disease by physician, John Marten, published in 1711.[12] Marten's work had received little attention because the pertinent selection was just a part of a larger chapter on gleets, a symptom of gonorrhea. The anonymous *Onania*, in contrast, generated dozens of reprints and translations and served as the basis for the pivotal 1758 and 1760 works *Tentamen de Morbis ex Manus-*

trupatione and *Onanisme*, respectively, by French physician S. A. Tissot. The latter work was translated into English in 1832.[13] None of the ideas were particularly new but the *Zeitgeist* was now favorable for such ideas as they fed into the current 18th-century medical debates on "cleanliness." *Onania*, originally designed to sell a quack remedy, established a new concern with the physical consequences of masturbation, one of which was impotence, and these notions persisted for the next 200 years—even longer among non-medical settings. Influential physicians in the 1700s took up the theory of debility combined with the moral proscription against sex for non-reproductive purposes and wrote about the adverse consequences of masturbation.[14] Benjamin Rush, a prominent American physician for the asylum movement, combined the moral and medical message in his injunctions against the crime in his 1812 work *Medical Inquiries Upon the Diseases of the Mind*, the first American textbook on psychiatry.[15] Licentious behavior, including masturbation, would deplete the overall amount of energy, resulting in debility, which in turn would lead to impotence. For the 18th-century public, the descriptions of post-masturbation diseases in *Onania* met the credibility test by an authoritative source and fitting the known anatomy and physiology of the day.[16]

The outcomes of this heinous crime ranged from desiccation of the vital organ to blindness and incorporated all manner of genito-urinary problems.[17] The central principle was that masturbation caused a debilitation (i.e., loss of fluid) more serious than that experienced with marital sexual activity.[18] In 1853, Claude-Francois Lallemand expanded in *A Practical Treatise* the debility theory to discuss spermatorrhea, a disease state characterized by loss of semen. Semen was thought to be represented by a fixed amount. Impotence was therefore an inevitable consequence of overuse, whether by masturbation or extra-marital intercourse.[19] It was an odd combination of morality and determined medical viewpoints. Painful urination, persistent erection, and "genital efflux" was due to the "weakening of the fibers of the genital organs and by the excessive relaxation of the pores and glandular orifices, which followed from the weakened fibers."[20] It was the latter effect that set the stage for Victorian medical professionals to incorporate these notions into their contemporary theories of mental illness.

The Victorian era was an "Age of Anxiety" and masturbation represented a "high water mark" of the perceived epidemic of anxiety-related nervous diseases.[21] Hall states that during this time, scaremongering about masturbation was thought to have done more harm than the actual practice.[22] Masturbation was still considered a sin according to both Catholic and Protestant church teachings because of the sexual behavior not being directed towards reproduction. The common view maintained that masturbation was a gateway to "fornication, disease and death, eroding self-discipline and self-control ... prostitution, debauchery, and obsession."[23] However, while even Hippocrates wrote about the possible effect of impotence from too much sex, physicians had not explicitly linked impotence with masturbation until now. The seeming contradiction of effects of too much sex within marital confines versus masturbation was solved because the latter was unnatural and "more subversive of the constitution."[24]

The Victorian medical model of neurology focused on nerve energy and self-pollution or masturbation was an obvious cause of depletion of that fixed amount of nerve force.[25] Tissot's manuscript, published almost a century earlier, had originally outlined the process of nervous expenditure during sex, noting that during intercourse with a partner, the sexual depletion of energy was "compensated by the magnetism of the partner."[26] In the 18th century and early decades of the 19th century, the concern for masturbation was possible congestion of blood flow to the brain, having been redirected from the sexual organs; by the 1850s, the energy depletion theory dominated medical discourse.[27] Thus, the excitation caused by masturbation resulted in a marked debility in the nervous system with many possible adverse consequences.

Concomitant with the nervous energy theory was a "hydraulic" model of male sexual functioning—a model we see today in a somewhat different guise with Viagra.[28] However, in the 19th century, there was thought to be a build-up of sexual energy or pressure requiring release, rather than redirection of blood flow. Nocturnal emissions were evidence of such build-up but more frequent emissions, such as those resulting from masturbation, were evidence of something far more serious and indicative of neurological dysfunction. If unchecked, this nerve deficit could result in the condition of sexual neurasthenia (i.e., sexual nerve weakness), which unchecked could ultimately develop into moral insanity. In fact, many lunatic asylum registries list masturbation as the cause of insanity among their male patients.[29] Conrad and Schneider call this period in Victorian medicine "the age of masturbatory insanity."[30] What was once a vice or sin was now becoming medicalized as illness.

Sexual neurasthenia was virtually synonymous with male hysteria during the late Victorian era. It was a variant of the neurological disorder, neurasthenia, now a discarded diagnosis but a very popular one during the late Victorian age. Neurasthenia was basically nervous exhaustion and much more about this disorder is discussed in another chapter. According to George Beard in 1900, the physician and leading expert on neurasthenia, sexual neurasthenia did not lead to insanity per se but some of the consequences of the primary disorder, such as masturbation and impotence, could do so.[31] He theorized that the reproductive system is the most primitive in the body with respect to evolution and so would be the first to be affected by disturbances of the nervous system. Men with strong constitutions can "bear the habit [of masturbation] without the warning effects on the nervous system that the sensitive feel, and if it is kept up year after year, insanity may follow."[32] He goes on to note that masturbation is an "effect as well as cause of insanity; that in this respect ... the tendency of insanity is to sweep away a moral sense and self-respect, and leave its victim an animal, and that one of the manifestations of this unchecked animalism is indulgence in this unnatural form."[33] Interestingly, Beard views impotence as a "disease of the imagination" and discusses the psychological elements and the role of ignorance of normal sexual functioning.[34] "Why is it," he asks, "that the slightest disease or suspicion of disease of the genital apparatus causes such absurd and unnecessary mental depression?"[35] The answer for him is the sympathetic nervous system with its connections to the brain and the stomach. "It is partly for this reason that sexual disorders so often excite neurasthenia or nervous exhaustion."[36]

While it was the female version that ultimately became infamous through the writings of Sigmund Freud and his mentor, Josef Breuer, male hysteria was initially considered to be the more common form of the disorder.[37] Famous men who suffered from neurasthenia included Theodore Roosevelt, brothers William and Henry James, and W. E. B. Du Bois. Social Darwinism and gender differences both had a role to play in the discourse of masturbation, neurasthenia, and moral insanity.[38] For men, the problem ultimately was a lack of self-control but note that the implication is that men *could* potentially have control over their sexuality. Masturbation was considered a disease of the Will wherein the compulsion towards onanism resulted in insanity and insanity produced onanism.[39] In contrast, women were perceived to be driven by their sexual organs. The classic Victorian gender division of active masculinity and passive femininity prevailed and influenced their respective treatments.

Likewise, some subgroups of males had "arrested development," as manifested by their inability to control sexual instincts.[40] For example, both manual laborers and Africans were thought to be immune to impotence due to their lesser brain endowments. "Only the White male had impotence and the self-control to overcome it."[41] Conversely, Victorian designations of medical deviance focused much on diseases of "Will," such that compulsions, including masturbation, led to disease.[42] Sexual neurasthenia and impotence were direct consequences of the environment of overcivilization and thus "brain-workers" were the most susceptible.[43]

The link with masturbation and insanity was a circular one: excessive masturbation caused insanity and yet excessive masturbation was a well-known consequence of insanity. Evidence was obtained by case study method and observations of mental asylum patients. Males were afflicted at rates triple that of females and the leading cause of male sexual neurasthenia (a milder disorder) was undoubtedly masturbation, at least in the minds of physicians. Impotence was an unfortunate consequence, no longer a divine curse but due to "overcivilization," one of the root causes of neurasthenia.[44]

William Acton's *The Function and Disorders of the Reproductive Organs*, published in 1857, was considered by both physicians and the public to be the authoritative source on the relationship of masturbation and impotence.[45] In the text, Acton references Lallemand's theory that nervous debilitation was a result of masturbation. However, Victorian physicians were not uniform in their attitudes towards impotence and masturbation.[46] One of the leading British physicians of the day, Sir James Paget, distinguished the moral from the medical issue and believed moderate levels of masturbation were normal; however, impotence was still considered a symptom of nervous system disease that could be brought upon by excessive loss of nerve energy through masturbation. Indeed, Paget viewed Acton as a nonprofessional quack (note that Acton was, in fact, not a physician but rather a "specialist") and publicly chastised his fellow physicians (Lallemand in particular) for creating needless anxiety among patients about their habit—an anxiety which only worsened the problem of impotence.[47]

The late Victorian era was the "high water mark of anxiety over masturbation."[48] Masturbation was not only a moral sin and resulted in general "depletion of health," but was a "gateway to fornication, disease and death, eroding self-discipline and self-control."[49] Even larger societal problems, such as prostitution and addiction were blamed on the practice. Not surprisingly, quackery was omnipresent, promising wide varieties of cures for both masturbation itself and its supposed consequence of impotence.

This was during this same time period that the "Purity" movement was gaining strength, as manifested in the creation of benevolent societies such as the White Cross League and Alliance of Honor.[50] This is not to be confused with the contemporary Purity movement among evangelical Christians although they have some similarities. The 19th-century movement's goal was primarily focused on chastity and the abolition of prostitution and other sex-related vices but a corollary principle was protection of the young from the practice of masturbation. Concern with masturbation during adolescence that could have negative consequences, such as impotence, in young adulthood sparked parental anxieties. In the 1890s, advice books for parents and children abounded amidst debates of whether boys reading such books might be influenced to engage in the sin rather than being kept in ignorance. *True Manliness*, a book by Jane Ellice Hopkins containing admonitions against masturbation, sold over a million copies by 1909.[51] In 1897 Sylvanus Stall had published a popular text, *What a Young Boy Ought to Know*.[52] In it he stated that masturbation could potentially lead to "idiocy ... early decline and death ... consumption ... total mental and physical self-destruction ... and, if one lived to manhood and marriage, could lead to inferior offspring."[53] Even the founder of the Boy Scouts, Robert Baden-Powell, in his famed 1908 *Scouting for Boys* addressed the issue in an appendix titled "Continence" noting that masturbation would lead to "weakness of head and heart and, if persisted in, idiocy and lunacy."[54] The references to the problems of "evil companions" among adolescent boys in school were related to the view that such behaviors could be contagious, i.e., the contagion of the habit of masturbation—not homosexuality, as one might assume.

By the last decade of the 19th century, one sees an emerging debate on whether masturbation is the cause or effect of insanity and the beliefs start to shift to the latter.[55] Havelock Ellis began publishing his various works on sexology in which he discredited some of the exaggerated claims and outlandish myths surrounding masturbation.[56] Instead he focused more on the negative effects of the persistent habit among those men of a certain psychological disposition. However, vestiges of the old attitudes remained in the early 20th century in the influential writings of psychologist, G. Stanley Hall on adolescence. A distinct picture emerged with the advent and popularization of Freudian psychiatry. In writings of Charcot, Janet, Freud, and Krafft-Ebing, they all supported the causal relationship of masturbation and neurosis, a term now beginning to supplant neurasthenia.[57] The physiological explanation was muddied, although still based on energy resources, but the dominant theme began to shift to the psychological. In the medical world, impotence fell squarely in the psychiatric camp—specifically, the psychoanalytic school of thought. It was no longer due directly to the physical act of masturbation but rather to the guilt and anxiety associated with the act in the practitioner himself.

The theory of nerve energy depletion had consequently given rise to the drug treatments for impotence. Whereas ancient and medieval remedies were confined largely to foods and herbs, Victorian drug therapy consisted of stimulants, such as strychnine and cantharidan in order to replenish the nervous energy.[58] Cantharides, extracted from the Spanish beetle, was the ancient aphrodisiac Spanish Fly. It was a topical irritant, however, and could result in blistering of the skin. Because opiates were at that time still considered stimulants, they were used by some with reported success in the middle decades of the Victorian age. But eventually, larger doses of opiates that came with chronic use and drug tolerance resulted in impotence from the drug itself. Cocaine and cannabis were potential treatments in small doses but for a limited time.

Non-pharmacological treatments, such as electricity, could also be applied to the groin's skin surface to restore energy. This restoration could be achieved by standard electrical stimulation or from electrical belts, manufactured by enthusiastic entrepreneurs. Physical exercise or exertion by manual labor was also recommended for males as a method to boost nervous energy and treat neurasthenia.[59]

A short-lived pharmaceutical treatment emerged during this time stemming from the Rejuvenation movement.[60] This movement developed from the research in the 1880s by neurologist and endocrinologist Eduoard Brown-Séquard, who experimented with injecting himself with extracts of animal testicular tissue as an anti-aging regimen.[61] The therapy had a potential side effect of treating impotence. The extract, made typically from the testicles of guinea pigs or dogs, became variously known as the Brown-Séquard Elixir or Pohl's Spermine Preparations.[62] This so-called "glandular solution" to aging was also derived from the energy model of the nervous system in that the "spermatic economy" had a fixed amount of energy and excessive spermatorrhea or masturbation could deplete that energy; however, if one could increase the fluid volume by injecting sperm material, then restoration could be complete.[63]

The idea behind rejuvenation more broadly was that one could prevent aging (and as a byproduct, enhance sexual performance) by injecting testicular extracts.[64] Eventually surgical treatments, essentially vasectomies, were developed in Europe in the 1920s in order to increase a buildup of sex cells. This presumably restored one's youthful prowess and masculinity.[65] The Steinach procedure, named after Eugen Steinach, the rejuvenator physician who developed the method, consisted of tying a ligature around the vas deferens to redirect sperm. The procedures were marketed in such a way as to play on the American national identity characterized by youthful energy. However, the practice fell out of favor, especially in the United States, as the continental practitioners were viewed with suspicion by the American medical establishment because of perceived unprofessional advertising and the stereotypically loose sexual mores of Europe. Those doctors engaged in practices we would now call direct-to-consumer marketing. In contrast to contemporary views, enhancement of sexual prowess was considered unnatural and inconsistent with normal aging.[66] As many liked to put it, "impotence was still largely regarded as a disease of the young, and a condition of the old."[67]

An examination of urology textbooks from the 1900s to the 1930s also reveals the gradual shift from a physiological explanation for impotence to a psychological or psychiatric one. The textbooks in the first decade of the 1900s still cling to the physiological relationship between genitalia and central nervous system using the diagnostic category of sexual neurasthenia.[68] A "temporary exhaustion of the nerve cells" is the fundamental root of the disorder, with masturbation as the most injurious precipitants. Admonition against masturbation is prominent, especially towards those with an inherited disposition to "neurotic taint."[69] According to Victorian notions of degeneracy, insanity could develop as a result of diasthesis—a combination of inheritance and behavior. Treatments laid out in these textbooks are typical of the late Victorian era mentioned above: electricity, stimulating baths, cantharides, phosphorus tonics, "nerviness," and a new drug—Yohimbine. Yohimbine is still sold today as a nutritional supplement that promotes weight loss, enhances muscle mass, and sexual desire despite little scientific evidence of its efficacy.[70] The urology textbooks in the teens had similar verbiage and advice, still holding on to the nerve energy model to explain impotence.[71]

However, a significant change is seen in the textbooks of the 1920s. Sexual neurasthenia has largely faded from medical discourse and is barely mentioned; furthermore, now there is no longer any reference to the nervous energy model. Instead, Freudian psychoanalytic causes are proposed along with the concomitant suggested treatment of psychoanalysis.[72] In another decade, the energy model of the nervous system had totally waned in popularity, although in the 1930s neurasthenia is sometimes listed as a neurosis causing impotence.[73] Impotence was by then universally viewed by physicians as a symptom of psychological sexual repression and subjected to psychoanalytic treatments. A few textbook authors allude to yohimbine as drug therapy but consider it to be limited in effectiveness.[74]

Lowsley and Kirwin, writing in 1940 were dour in their prognosis for treating impotence: "The prognosis is bad, and treatment of little use, when the impotence is dependent upon organic cerebrospinal lesions ... [when purely functional, it is] a neurological rather than a urological problem."[75] Interestingly, the reference to organic versus functional disorder is a harkening back to 19th-century neurology. They then went on to prescribe complete abstinence "so as to give the exhausted nerve centers a chance to renew themselves."[76] The rhetoric is a complete throw-back to Victorian views.

The last reference to neurasthenia as a contributor to impotence is found in 1948 and from then on, if an explanation for the problem is even given, it is only a psychiatric one.[77] The commonly cited figure of 95 percent of impotence cases being due to emotional factors first appears in the 1950s textbooks and is then repeated in subsequent textbook editions up through contemporary times.[78] One interesting treatment is mentioned by Parton in 1960 in his medical book geared towards general practitioners.[79] He describes using a placebo approach of prescribing nine white-colored pills and one colored pill. The white pills were to be taken once a day, 15 minutes prior to bedtime. On "la nuit d'amour," the colored pill should be taken.[80] Parton indicates that the doctor must give the patient a good "sales talk" about the procedure.

Whereas 20th-century medical therapies for impotence included some approaches that focused on mechanics of augmentation such as vacuum pumps, treatment had become almost entirely the provenance of psychiatry. For most of the century, impotence in the younger male population remained a psychological disorder with no known physical cause and was even listed in the first edition of the *Diagnostic & Statistical Manual* (*DSM*), the primary classification scheme for psychiatric disorders by the American Psychiatric Association.[81] In the first half of the century, blame was put upon the so-called "frigid wife."[82] This explanation fell out of favor in clinical psychology with the rise of behaviorism as the ruling school of thought and the declining of psychoanalytic theory among practitioners. Yet no serious attempts at non-psychological therapies were made until the waning years of the 20th century.

In the 1990s urologists reclaimed sexual dysfunction as a medical disorder in a transfer of "expert ownership."[83] This change in power was in part due to the urologists seeking new revenue sources and resulted in new emphases first on surgeries and then ultimately, on "lifestyle" drugs. The discovery of the vasodilator, sildenafil, clinched the transformation. Impotence became simply a technical problem with an easy fix and "the medical model eclipsed the psychogenic model."[84] Sildenafil was branded Viagra and with de-regulation of Big Pharma during the Reagan administration, was directly marketed to the American male public; likewise, mild impotence was rebranded Erectile Dysfunction and was enlarged as a diagnostic category to include *any* dissatisfaction with one's sex life: this redefinition had the effect of tripling the incidence of impotence cases within three years. Here the pharmaceutical company was behind the creation and labeling of erectile dysfunction with a motivation of profits. Only *after* the successful launch of Viagra, did the scientific community finally solve the mystery of the physiology of an erection. The drug's success had already been established despite failure to understand the causes of the disorder, impotence, and even then, the drug was shown to have a 40 percent placebo rate. Placebo rates this high typically indicate some psychological component to a condition.

Given the great pharmaceutical success of Viagra for the treatment of male impotence, is there a similar remedy for the female equivalent? Female Sexual Dysfunction, or the more specific Hypoactive Sexual Desire Disorder, is another newfound disorder in search of a remedy. The simple assertion in 1999 by a group of urologists, most of whom had ties to the pharmaceutical industry, that such a disorder existed as a biomedical phenomenon led to the disorder's widespread acceptance first by the medical community and then the general public in the early years of the 21st century. Like impotence, frigidity had been added as a disorder in the American Psychiatric Association's *Diagnostic and Statistical Manual* in 1952.[85] Now portrayed as a medical disorder, Female Sexual Dysfunction clearly needed a medicinal remedy.[86] Flibanseran (Addyi) was eventually approved in 2015 by the FDA as the first pharmaceutical to treat the disorder. Unlike sildenafil, flibanseran directly affects the brain by affecting the neurotransmitters serotonin, dopamine, and norepinephrine. The drug had originally been studied as an antidepressant and essentially its formulation is typical of that class of drugs and so works by the same neurotransmitters implicated in depression.

The FDA process was not smooth. There was significant debate on the medicalization of women's sexuality, efficacy of the drug, and whether the significant negative side effects were considered acceptable, given the small positive effect of the drug. Twice rejected, final approval by the FDA occurred after major lobbying efforts by the manufacturer in the face of a tepid benefit/risk assessments. The effectiveness of flibanseran was found to be only marginal and barely above effects seen with placebos; conversely, the probability of adverse side effects was elevated.[87]

Forbes reported on how successful Addyi had been a year after approval in 2016. At that time, sales had not yet met expectations. Demand by patients had not materialized for several possible reasons. Taking a daily medication (versus sildenafil or equivalent only taken as needed) that acted on brain chemistry for a problem whose medical origin was suspect apparently held little appeal. The negative side effects likely had an effect on demand and compliance as well.

Thus, we have a tale of social construction of illness that began with what was formerly considered to be a normal consequence of aging. Viagra was a "solution in search of a problem" and its spectacular financial success is due not just because of the pharmacokinetics—it does indeed cause erections—but because of the transformation of the illness narrative.[88] As one of the first lifestyle drugs to be legally mass-marketed directly to the public, Viagra enjoys almost unrivalled brand-name recognition. But the story of Viagra is also a unique neuroscience story in that the so-called disorder of impotence or erectile dysfunction had its etiology founded firstly in sin, secondly in a medical model of physiology interacting with a moral model that still included notions of sin or vice, then a psychological model fed by guilt, and now back to a purely mechanical medical model. The common threads throughout these different iterations prior to the advent of Viagra have been the perceived interaction of mind and body and which aspect enjoys the expertise and hegemony over the condition of impotence. That expertise has variously been associated with religious authorities, apothecaries and quacks, physicians, psychiatrists, sexologists, and now pharmaceutical companies. The problem of impotence was originally seen as a spiritual one and now is totally relegated to the provenance of the body. In Victorian times, its historical link with the heinous sin of self-pollution and moral insanity was mediated by the nervous system. In contemporary times, because treatment using sildenafil only addresses the "hydraulics" with little nervous system influence, the psychological factors once thought vital to the understanding of impotence in younger men have gone by the wayside. Instead of a typical consequence of aging, impotence is now seen as abnormal in the elderly and yet another condition worthy of pharmacotherapy.

10. What's Your Type? Typologies

GIVEN THAT THERE IS NOW a sub-discipline of or research on the neuroscience of just about every human behavior from sleep to sex, it should hardly be surprising to learn that there is now a subfield of personality neuroscience.[1] This area of neuroscience focuses on the biological basis of personality. At its roots, however, stand the ancient Hellenic traditions in medicine that attributed temperament to both physical and physiological processes. The modern neurobiological perspective may now center on individual differences in nervous system physiology, namely specific neurotransmitters and brain regions or structures, but the core principles underlying the theories remain largely the same. While humans have in common the basic structure and function of nervous tissue and systems, they also have variation among them which represents either categories or continua of individual differences. The distinction between the discourse of modern neuroscience and historical times is one of details rather than general substance of the theories. An interesting feature of the history of personality psychology is that in the discipline's development, there were two separate and independent strands of intellectual knowledge that were woven in parallel for decades; one was from the medical or clinical tradition and the other from the scientific world of experimental psychology.

Temperament is a term that has been consistently applied throughout the ages to examine individual differences whereas personality is a relatively modern term, used by the scientific community for only the last one hundred years or so when testing and measurements became a popular application of the new psychology.[2] Temperament, however, has long been a descriptor of the human character. The most common and influential descriptive scheme for temperament was the ancient Greek humoral doctrine of Galen, later adopted by Hippocrates.

While hardly scientific by today's standards, the humoral doctrine was based on the physiological theory of the day. It was widely believed that *spirits* mediated between the material of the body and the ether of the mind. They were the links between mind and body through which nerves and muscles were innervated. According to Smith, spirits made the potential of the body become actual with sensation and movement, for example.[3] The material body was associated with the four chief elements of the physical world: air, water, earth, and fire. These elements in turn were then later associated in Hippocratic medicine with the four key bodily substances—black bile, phlegm, yellow bile, and blood—and now corresponding with the four humors—cold and dry, cold and moist, hot and dry, hot and moist, respectively.[4] From there a theory of temperament emerged that would dominate medical, philosophical, and popular discourse for many centuries. Depending on the balance of the humors or, more likely, the relative dominance of one of the humors, an individual's temperament might be melancholic, phlegmatic, choleric, or sanguine. Physiological humors corresponded to basic temperaments or types. Well into the 1800s, physicians practiced a medicine that was firmly rooted in the humor doctrine in which optimal health was a function of balance among the four humors. Techniques included blood-letting and cupping, for example, to restore relative balance.

So here the link between temperament or personality was explicitly tied to what was then contemporary physiological theories underlying nervous system function. The doctrine of the humors accounted for a biological basis of individual differences that meshed with the integrative view of humans and the cosmos. As seen in expositions of psychopathologies, the theory not only dealt with the causal factors but provided grounds for predictions and applications for therapeutic purposes. The resulting typology, while simple and straightforward, was an appealing categorical framework to peg one's character and understand individual temperaments. Furthermore, it had the backing of science—such as it was understood back in the day.

As scientists began to home in on the brain and nervous system being the physical repository of one's personality as opposed to the various substances of the body whole, new paradigms emerged. George Rousseau argues that the shift from a humoral doctrine to one of temperament defined by "nervous physiology" was accomplished in the 18th century.[5] The discoveries regarding the circulatory and nervous systems by William

Harvey and Thomas Willis, respectively, in the 1600s precipitated that theoretical shift in scientific discourse. Willis had shown that the brain and nerves, rather than humors, had to be involved in neuro-pathologies. By the dawn of the 19th century, there was consensus that temperament was formed through the interaction of nervous physiology and experience. Rousseau, in fact, uses this historical analysis to illustrate the intellectual heritage of one of the prominent contemporary theories of temperament, that of Jerome Kagan (to be discussed later in this chapter).[6] Much as Kagan does, 18th-century practitioners examined temperament with respect to socio-cultural factors such as race, gender, and class.[7] However, was the theory of nervous physiology developed and interpreted in ways to confirm societal views of temperament in respect to maintenance of the social order rather than challenging the status quo? The same question could be asked today, as Kagan suggests.[8]

Phrenology, the popular pseudoscience in the 19th century, capitalized on the new anatomical evidence regarding the brain's role in governing behavior. While the theory was ultimately disproven and abandoned by mainstream physicians and physiologists, it provided the impetus for a lively debate of distribution of mental faculties versus localization. This debate still goes on today in neuroscience.

The concept of mental faculties is not equivalent to temperament or personality but there was nonetheless some overlapping of constructs. Phrenologists also spoke of "propensities," aspects of character which very much were considered to reflect one's personality or temperament.[9] The propensities of "amativeness" and "secretiveness" were especially distinguished to differentiate individual characteristics.

As discussed at length in Chapter 1, when phrenology first appeared on the scientific scene, it was largely viewed as representing the most advanced understanding of the workings of the nervous system. Its founder, Franz Joseph Gall, was a renowned anatomist and indeed today gets credit for settling the debate once and for all on the brain being the seat of the mind and establishing the route of nerve tracts from the spinal cord to their points of origin or terminus in the brain by utilizing a novel dissection technique. Phrenology was considered the "new science of the mind, an empirically based social philosophy and an infallible guide to individual character."[10] Some historians view Gall's work as an important shift from placing psychology (and, consequently, personality) from the philosophical realm to the physiological one.[11]

The premise of phrenology was that the brain is made up of distinct organs corresponding to individual faculties divided among propensities, sentiments, and intellectual abilities rather than a "common sensorium."[12] Enhanced or diminished activity of one of those organs was reflected by greater or less physical volume of brain tissue, respectively, devoted to the organ. Since the skull was thought to be a faithful representative of the cortical tissue below, one could obtain a character profile of an individual simply by measuring the protuberances of the head; cranioscopy allowed for a non-invasive examination of the brain. This feature of the theory proved ultimately to be a major factor in phrenology's downfall as anatomists finally demonstrated that there was little correlation between the shape of the brain and the bumps on the skull. The key point here, regardless that the fate of phrenology to become relegated to the heap of pseudoscience, is that personality or character was revealed by external features that were related to internal anatomy and physiology.

A similar assumption had been made in the case of physiognomy. This was an earlier attempt at promoting a physical basis for temperament. Again, using the premise that external physical features revealed internal personality characteristics, physiognomy was centered on facial features and expressions. It was a method of discerning character from one's expressions.[13] While phrenology enjoyed a relatively brief popularity among the scientific community between the 1790s and 1830s (with a popular resurgence during the late Victorian era), physiognomic theory was first popularized in the 16th century and persisted until the early 19th century. According to Parssinen, physiognomy "paved the way" for phrenology with its emphasis on the forehead in facial expression.[14] The nomenclature of the organs in phrenology had its origins in physiognomy.[15] Johannes Spurzheim, Gall's assistant and later chief promoter of phrenology, published in 1815 a work entitled "The Physiognomical System of Drs. Gall and Spurzheim," which indicates the linkage between the two approaches.[16]

Physiognomy was first introduced by Giovanni Battista della Porta (1536–1605) but it was most associated with the work of Johann Lavater (1741–1801), whose silhouettes represented keys to reading character in people's faces.[17] It was an early study in body language—a topic that continues to fascinate the public today—and provided a "language that domesticated the soul."[18] Physiognomy "offered a spiritual guarantee that anyone could read the appearances of things in the world and then form a judgement on the basis of their essential yet hidden value."[19] Its practice was built on the observation that people do indeed make judgments about others based on appearances and so Lavater attempted to analyze and systematize that process. Of course, the judgments were heavily biased by sociocultural assumptions that reflected a status quo regarding economic power and other hegemonies. Lavater nonetheless believed that the lines of the face and expressions wrought by one's passions revealed the character of the "hidden man."[20]

Similar to the case of phrenology, practitioners of physiognomy held the view "that which we take to be peculiar and distinctive to an individual then becomes that which is common and ordinary to all individuals."[21] Because one can create broad classifications of categories to generalize personality characteristics, the obverse of that premise is that individual differences among those characteristics can be readily discerned.

Where the two practices differed was in the religious implications of each theory. Phrenology challenged the religious tenets regarding an immaterial soul; the charge of materialism of the mind was serious enough to exile Gall from his native land for fear of prosecution of heresy by the Church.[22] A physiological basis for the mind held no place in Church doctrine in the 18th century. Thus, Gall had to perform some semantic convolutions in crafting a theory that would remain consistent with the prevailing views, namely, by creating an organ in the brain for knowing God. In contrast, physiognomy had no such conflict with religious views and indeed was a vehicle to maintain and support the morality of the times. While the theory was considered a "science of the mind," facial expression was a product of the mind and the mind reflected God.[23] Only the divine presence could line the face in such a way to promote the dignity of man and reveal the class and race distinctions that ordered society. It was nature's (i.e., God's) own sign system.[24] The morality expressed represented widely held beliefs about culture and society and physiognomy became a popular means of categorizing and describing people.

Physiognomy was a typology in the sense that it involved "comparison and classification, seeing the general in the particular and the universal in the general."[25] By observing expressions carefully, one could eventually distinguish between the "original and habitual expressions from accidental ones."[26] Complex expressions were reducible to simple or "typical" ones and consequently were analyzed into standardized types. This was one of the attractions of both phrenology and physiognomy: people could be easily trained to be expert observers of external features of the body, whether examining skull contours directly from the brain or facial features as a proxy of the brain. Someone's countenance could mask inner feelings or personality but the anatomical structure—the angle of chin and nose, the forehead height, and so forth—could not be hidden.[27] A readily accessible system based on standardized observation allowed people to analyze and reduce the infinite variety of facial expressions to a simpler, easily generalizable type.

They were systematic theories that allowed common interpretations to reflect socio-cultural norms. Categories generated essential types that followed the tri-partite format promulgated by Plato, Church philosophers, and later even Sigmund Freud: animal, intellectual, and moral (i.e., sense, reason, and will; id, ego, and superego). The moral sense was the unifying feature between mind and body and distinguished humans from lower animals. Lavater believed that facial expression reflected the mind and hence the soul; therefore, it was indicative of an individual's character.[28]

The other point of similarity of phrenology and physiognomy came with their respective practical applications in society and broad acceptance in popular culture as reliable indicators of personality. Long after science had abandoned both fields, the practices retained their hold on the public as reliable methods of ascertaining character and temperament and consequently, suitability for specific vocations, marriage partners, and propensity towards crime. Physiognomy held sway among Victorians, especially when bolstered by Darwin's evolutionary theory, in promoting and maintaining views regarding ethnicity and race. Its tenets became justification for beliefs regarding the relative superiority or inferiority of a particular race, for example. Lavater in the late 18th century wrote of national characteristics to base generalizations of both physical features and differences in temperament.[29] Later principles of degeneracy and eugenics built upon the earlier concepts of physiognomy.[30]

But finally, at their core, both approaches reflected views of how the mind influenced the body. The former humors or bodily fluids that governed one's temperament were gradually being supplanted by the brain in a paradigm shift from the 18th to 19th centuries.[31] Phrenology had placed the brain at the center of the mind and even physiognomy was thought to reflect a "nervous sympathy between the diaphragm" and the "transmission of impressions along the nerves to the face."[32] Both theories were based on physical principles as known at the time to link biology and the mind. Marshall Hall's view of the nervous system was beginning to take hold in the 1830s with his notion of separate entities.[33] There was the cerebral nervous system (the brain), the spinal nervous system, and a ganglionic system. The latter two were the source of vegetative functions, reflexes, and the passions; in other words, they were the seats of involuntary action in contrast to the brain which housed our intellect and reason.[34] While Hall's delineation would eventually be discarded by neurophysiologists, his further explication of the reflex arc represented a significant advance in the understanding of nervous system function. Along with the advent of Darwin's theory of natural selection in the mid–19th century, the spiritual entwinement with the nervous system was gradually phasing out of scientific discourse. The idea that character can be observed from visual appearances—whether from the face or the contours of the brain—begins to wane as the source of character becomes more diffused or ambiguous rather than being centralized.[35]

The early decades of the 20th century brought the next attempts at a personality typology based upon principles of neurophysiology. Edward Kempf proposed a type theory in 1918 based on autonomic functions of hormones.[36] Here he proposed six personality types based on the relative strength or weakness of the autonomic nervous system. It was a curious application of the current research on the physiology of emotion, comparing the respective James-Lange and Cannon-Bard theories (both of which focused on the role of the sympathetic nervous system and perception), to Freudian theories of wish fulfillment and gratification of the Id. Neither this neurohormonal theory nor a more popular concurrent one proposed by Louis Berman in 1921 achieved much long-lasting traction.[37] Berman's typology was based purely upon the endocrine system, dividing types among the major hormonal gland systems of the body: adrenal, pituitary, gonads, thyroid, and thymus. Besides the physiological issue of the pituitary gland influencing most of these other endocrine glands and thus having considerable redundancy, even Berman's own work found that most people were of mixed types. And so, the hormonal approach was largely abandoned.

The more influential typology during this same time period came from a seemingly unlikely source: Ivan Pavlov (1849–1936). While best known for his eponymous theory of classical conditioning in which learning is achieved by associating reflex action with neutral stimuli (e.g., salivation to the sound of a bell), Pavlov was more interested in developing a comprehensive and integrative theory of higher nervous system action.[38] His British contemporary Charles Sherrington (1904) pursued a similar route of physiological research and viewed the reflex arc in a functional light, showing how the sensory and motor elements could be either excited or inhibited by the brain.[39] His work, however, was more confined to explicating the individual neurons involved in complex reflexes and showing how they operated as a functional unit, sometimes independently but often reciprocally with other neural circuits. In contrast, Pavlov's research utilized his learning paradigm to found a new unified science regarding higher level functioning of the brain.[40]

The application of Pavlovian theory to personality typologies came from his notions of excitation and inhibition in the brain.[41] The types or classifications of temperaments came from aspects of neural conditioning: the speed of neural processing, the proportion or balance of inhibition and excitation, and the relative strength of inhibition/excitation. The possible types, based on these three dimensions of cortical functioning were "(a) strong, balanced, and mobile (sanguine); (b) strong, balanced, and slow (phlegm); (c) strong, unbalanced, and impetuous (choleric); and (d) weak (melancholic)."[42] The reference to the original four humors should not go unnoticed! While Pavlov's research was conducted on animals, he was keen to generalize his notions of inhibition and excitation to human brain function both in the areas of personality and mental health.[43]

Pavlov conceived of inhibition and excitation as general processes that could work reciprocally by inducing the opposite effects within cortical fields. Nervous centers of excitation or inhibition "irradiated" and interacted with one another to produce an equilibrium state.[44] Whereas balance, strength, and lability of inhibition and excitation formed the basis of temperament, a state of disequilibrium could result in psychopathologies. Pavlov's typology did not receive the same attention in the United States and Western Europe as did his tool for studying it—classical conditioning. Instead, when discussing neural integration at the cortical level, neurophysiologists and psychologists focused on Sherrington's synaptic views of neuronal connections. It was a micro rather than a macro approach that had little to do with temperament.

However, the German-British psychologist, Hans Eysenck, utilized Pavlov's theory to undergird his own attempt at a biologically-based typology for personality.[45] Coming from a tradition of studying individual differences using quantitative methodologies, Eysenck developed a three-factor theory of personality. As the behaviorist perspective dominated Western psychology during the 1950s, Pavlov's typology theory was attractive to Eysenck in that it provided a biological basis for individual differences that could be tested empirically.[46] Eysenck's three factors were dimensional axes rather than discrete categories and incorporated Pavlov's ideas of excitation and inhibition as explanatory mechanisms for the dimensions.[47] Specifically, the dimension with opposite poles of introversion and extraversion was conceived as a function of greater or lesser cortical arousal (excitation) and likewise, cortical control (inhibition). The factors of introversion and extraversion had been proposed decades earlier by Carl Jung as attitudes towards the world, whether being inwardly or outwardly focused.[48] While Jung's own typology theory was and continues to be popular (now disguised within the framework of the ubiquitous Myers-Briggs classification scheme), Jung did not emphasize a biological underpinning despite implications that there must be one. The other factors in Eysenck's theory were neurosis and psychosis, both reflecting the dominance of Freudian terminology and concepts at the time. The dimension of neurosis was anchored at the opposite pole by stability, also a potential characteristic of cortical function. The third axis, psychosis, was a continuum with impulse control at its opposite. Smith argues that Eysenck "restated in a richly suggestive manner the nineteenth-century language that related internal individual control to external social forces."[49]

Despite Eysenck's tutelage from some of the most renowned psychometricians with a lineage back to Francis Gall, his own background was from a clinical perspective. And so he developed clinical assessment instruments that would determine an individual's place in this three-dimensional psychological space and used that data to generate testable hypotheses about physiological responses and eventually about susceptibility to particular pathophysiological states, such as drug addiction and cancer.[50] For example, he predicted that because introverted individuals actually had greater levels of cortical excitation, they would be more likely to abuse nervous system depressants; conversely, extraverts with their lower levels of excitation and greater inhibition would be more likely to abuse nervous system stimulants. Other researchers confirmed with EEG studies his predictions of cortical arousal in relation to types.[51]

Eysenck's theory was unabashedly a typology with a clear biological foundation in the nervous system but it became somewhat buried in the deluge of trait theories that were emerging from experimental psychology during the same years. However, most of those trait theories, including the most popular one today—the Five Factor Theory—incorporate some of Eysenck's ideas, particularly those related to introversion and extraversion. It is likely, in fact, that this factor (or trait) has been the most thoroughly researched and, as discussed later in this chapter, still has the best empirical support for a biological rather than environmental origin.

Before moving on to contemporary research in personality neuroscience, one other historical approach of relating biology to personality should be examined. This is constitutional or somatotype theory, as exemplified by Ernst Kretschmer (1888–1964) in Germany and William Sheldon (1898–1977) in the United States. These theories were based on the premise that external body type was related to personality and, in Kretschmer's theory, proclivity to specific psychopathologies. In a sense, they were simply variations on the physiognomy theme but instead of just focusing on the face and head, incorporated a holistic approach utilizing the entire body shape. Both theorists proposed three somatotypes, essential body types that could be used as a global classification scheme. Kretschmer proposed the somatotypes of pyknic (fat), leptosome (thin), and athletic; Sheldon famously had the endomorph (fat), ectomorph (thin), and mesomorph (athletic; average) somatotypes.[52] While these somatotypes were related to temperament by their researchers, their popularity lay in their correlation with mental and physical health. Sheldon's constitutional theory was especially successful in capturing the public's attention and filtered into both common discourse ("the jolly fat man") and ideals for physical education. Being grounded in psychiatry, Kretschmer's work showed most success in demonstrating links with personalities prone to specific mental disorders (e.g., schizophrenia, mood disorders) but was considered narrow in scope and had largely disappeared into obsolescence by the 1940s.[53] Despite some success in demonstrating physiological differences in neural responding, his types were too simplistic to encompass the variations of mixed types seen in both normal and abnormal populations.[54] Sheldon's theory also declined in popularity because of its connection with eugenics and attendant racist implications. It was "a 20th-century physiognomy linking physique with temperament" and Sheldon was a self-proclaimed "biological determinist," who believed somatotype influenced intelligence, racial superiority or inferiority, and success in life.[55] However, somatotype theories ultimately suffered from a fatal research flaw of inferring causation from correlation. For example, criticisms of Kretschmer's typology grew with empirical evidence demonstrating the confounding of body type and psychiatric diagnosis with age.[56] With the rise of modern trait theories and their more sophisticated research designs, these simplistic constitutional theories were discarded and relegated to the heap of historical typologies.

Modern trait theories have biological components as well and indeed emphasize the role of nature over nurture. Their focus, however, is on genetic endowment and heritability and their interaction with learning and the environment. These approaches come predominantly from the experimental psychology field and are heavily based on psychometric methodologies using highly complex statistical techniques. They are not typologies in the traditional sense but there is overlap in some of their theoretical assumptions, namely, that temperament can be governed by biological individual differences. Gordon Allport, considered to be the "father of personality theories," defined traits as "neuro-psychic structures," clearly suggesting a component of neural functioning being an important element to personality.[57]

The most popular of these trait theories continues to be the Five Factor Model developed by McCrae and Costa in the 1980s, sometimes simply known as the Big Five.[58] This model posits five major traits: openness, conscientiousness, extraversion, agreeableness, and neuroticism. As the name suggests, it is not actually a theory per se but a model of five personality factors that consistently emerge from statistical factor analysis of character descriptors. The factors come from a lexical tradition, first espoused by Allport, in which personality adjectives are analyzed into clusters without regard for theoretical or causal mechanisms.[59] Each factor can be considered a bi-dimensional axis with polar opposites, e.g., extraversion versus introversion. While not resulting in a typology as such, with so few factors, the permutations do tend to be interpreted in ways similar to types. McCrae and Costa are not the only researchers to focus on these constructs either singly or as a cluster, but the sheer productivity of their empirical research and the continual generation of new research hypotheses over the last 30 or so years have lifted their particular model to a height of prominence over the others.

The model takes into account genetic predispositions along with character adaptations. John and Srivastava (1999), citing the work of Plomin and Caspi, state that personality dispositions must have a genetic basis and "must therefore derive, in part, from biological structures and processes, such as specific gene loci, brain regions (e.g., the amygdala), neurotransmitters (e.g., dopamine), hormones (e.g., testosterone), and so on."[60] This statement summarizes neatly the scope of contemporary personality neuroscience. Current research is focused on identifying brain regions via imaging techniques, neurotransmitters and hormonal responses, and molecular genetics that underlay individual differences with respect to the Big Five factors.

Examining heritability of personality traits has been a long-standing research program for the last 50 years. Early twin study research suggested that heritability ranged from percentages in the low 30s to a high around 70 percent.[61] However, the search for any single "personality" gene has been elusive and probably misguided. Setting aside the cultural, semantic, redundancy (non-orthogonal factors), and tautological issues by which labeling traits have been beset, finding a single gene or even portions of genes (SNPs or single nucleotide polymorphisms) for a particular personality trait has not been reliably demonstrated.[62]

A more promising approach has been focusing on smaller units of behavior thought to underlie a trait that relate to more fundamental animal behaviors (and thus lend themselves more readily to behavioral neuroscience research). Some personality researchers refer to these as lower order traits as opposed to the higher order traits or "superfactors."[63] These behaviors include reward-seeking, avoidance of threat or punishment, exploration/curiosity, attentional mechanisms, and emotional regulation.[64] By singling out these behavioral tendencies, specific neurotransmitters have been identified as mediators and likewise, specific brain regions have been associated.

The neurotransmitter dopamine has faced intense scrutiny in personality neuroscience. It has been implicated in the traits of extraversion and aggression, for example.[65] One of the many challenges in this type of research is that a single neurotransmitter is not only found in multiple pathways to many different brain structures or nuclei, it also has a multitude of different receptor sites. Depending on the receptor site, neural inhibition or neural excitation can occur and so a neurotransmitter can have a plethora of effects.[66] However, there are some emerging data zeroing in on the behavioral function of some of these pathways and the ways dopamine might be related to certain personality traits.

Three major dopamine pathways in the brain have been identified thus far: (1) nigrostriatal, dealing with the motor system; (2) mesolimbic, the reward pathway; and (3) mesocortical, concerned with higher cognitive functions.[67] One factor that links these three functions is extraversion, which could be considered the manifestation of reward-processing. The nigrostriatal pathway is involved with converting incentive motivation to a motor response, i.e., a reward-seeking behavior or strategy, while the mesocortical system deals with the attention and working memory necessary to attain goals. Individual differences in extraversion/introversion may lay in the dopamine receptor synaptic responses. Dopamine regulation may be the key to the trait or super factor of extraversion.

Similarly, brain volume and imaging studies have linked extraversion and pertinent brain regions associated with the reward pathway.[68] The nucleus accumbens, amygdala, and the medial orbitofrontal cortex, for example, have been implicated in these studies of sensitivity to and coding of reward. Links with other brain areas have been made individually with the other Big Five traits. The assumption of these studies is that if a "trait is hypothesized to be associated with the functions of a set of brain regions, the finding that the trait in question is associated with structural variation in any of those regions supports the hypothesis."[69] This finding would support the notion of biologically-determined individual differences. While a lot of research attention has been placed on the role of reward-seeking in extraversion, DeYoung suggests that the construct of "exploration" be considered the unifying factor of dopamine pathways.[70] This would account for the actions of different dopamine receptors and their respective pathways. In this paradigm, the dopamine system is viewed as a hierarchical system with exploration as the higher order function operating in the extravert. Only this type of holistic system can purportedly deal with opposing dopamine responses among different pathways. In this system, the three relevant characteristics of dopamine are the overall levels available, the activity level of dopamine "value coding" of reward, and the activity level of dopamine "salience coding" of reward.[71] In other words, how much dopamine is available, how attractive is the reward, and how important is the reward? Extraversion is most associated with value coding, according to DeYoung.

Hermes and colleagues provide similar empirical support in their research on extraversion.[72] They propose that positive affect is a core element of extraversion, perhaps the most important element, and utilizes the same mesolimbic pathways (e.g., ventral tegmental area, nucleus accumbens) for reward and incentive motivation discussed previously. Furthermore, they note that research shows that introverts and extraverts respond differently in choice reaction time tasks when given dopamine antagonists; likewise, dopamine agonists can affect their hormonal responses differentially. Finally, there are differences between introverts and extraverts to positive affect versus rest or neutral experimental conditions.

However, these temperamental differences should be viewed within a developmental framework of self-regulation.[73] As the mesolimbic pathway and areas develop and mature, attentional mechanisms and executive control features begin to regulate or modulate activity. While there are individual differences regarding temperament that influence an individual's reaction to environmental stimuli, there are distinctive interactions with cognitive processes. These patterns of self-regulation emerge later in childhood.

A newer approach in personality neuroscience has focused on connectivity within these proposed networks associated with specific traits. Rather than emphasizing a region, structure, or nucleus, examining the density and integrity of white matter (myelinated axons) can provide a basis for comparison of individual differences.[74] Here, the focus is searching for "networks of interest" (NOIs) in addition to "regions of interest" (ROIs) that have usually been identified prior to conducting the study.

What do these contemporary research trends have to do with the typologies of the past? There is a fairly straightforward path from Pavlov's rendition of excitatory versus inhibitory cerebral dominance to Eysenck's cortical arousal typology to the current dopamine activation theories. It is not too much of a stretch to see researchers in the future categorizing personalities into dopamine-dominant or norepinephrine dominant (as is posited with the trait of neuroticism) types. Instead of the humors of black or yellow bile, blood, and phlegm, we now have dopamine, norepinephrine, serotonin, and GABA (gamma-amino butyric acid). Increases or decreases of these chemicals correspond to different personality traits or types. Curiously, this approach is more similar to the hormonal ones of Kempf and Berman a century ago. A key problem with those theories was the interaction and interdependence of the hormones and that issue remains with neurotransmitters. For example, dopamine systems are impacted by glutamate systems, making it difficult to tease apart the pertinent effects.

But just like the Galenic humor doctrine, phrenology, and physiognomy before, the current biological approaches to personality are fraught with methodological and philosophical pitfalls. There exists the danger of interpreting results to fit a pre-existing narrative rather than systematically testing data against hypotheses. For example, DeYoung et al. note that while empirical evidence supports the hypothesis that greater than average brain region volume suggests greater than average function, the opposite finding might still support greater function because less volume could represent greater efficiencies from synaptic pruning.[75] In other words, either outcome can be made to fit the claim of a particular brain region being implicated in the personality trait under investigation. When all results, including contradictory ones, are assimilated into an existing theory rather than accommodating the theory to fit new facts, there is a danger of losing scientific credibility. When hypotheses are incapable of being falsified, a basic tenet of the scientific method has been violated. However, in the early phases of scientific investigation in a relatively young discipline, description of phenomena is often the first order of business and then followed by systematic, inferential hypothesis testing. Contradictory or unexpected findings can and should pave the way towards better hypothesis testing and more refined theorizing. There is genuine enthusiasm and optimism concerning the strides made in discerning the molecular genetics that govern personality characteristics and improvements on experimental design, task conditions, and assessment measures will continue to advance the field.[76]

Typologies are, by nature, also vulnerable to tautological explanations. This woman is an extravert because she is outgoing; this woman is outgoing because she is an extravert. How can one get out of the explanatory loop? The answer lays in utilizing the methodologies that assess construct validity. Using both convergent and divergent evidence against external constructs that are either related or quite different, researchers can establish the hypothetical validity of a trait.

But this then leads to another philosophical issue to be discussed: why types, why categories? There has been a long fascination with nominal classification schemes in Western science, whether making a taxonomy of the natural world or a nosology of psychiatric conditions. There is often an assumption of a categorical framework in which to place discrete items according to a particular criterion when the construct is actually an ordinal or dimensional framework. Typologies that are simplistic run the risk of being too broad to be useful in describing or predicting behavior. Yet if they are too complex, their utility as a short-hand communication tool is limited. It is also possible that the construct exists or operates under rules that are currently incomprehensible or even inconceivable in our current scientific schemas. Our current scientific epistemologies or ways of knowing may someday undergo a major paradigm shift that completely up-ends the notion of a typology.

Finally, does a neuroscientific basis of personality give us "value-added" information in understanding human behavior? Jerome Kagan, one of the most influential contemporary theorists and researchers in cognitive development, is perhaps best known for his work on temperament in the child.[77] Decades of research by Kagan point to the seminal role that biologically-based temperamental differences observed in infants play in adolescent and, subsequently, adult behaviors. However, he gives equal if not greater importance to role of psychological and socio-cultural factors, especially those that contribute to a child's identification with race, class, and gender. He applauds the current trends for combining biological and psychological variables yet cautions that biological factors have been overemphasized.[78] The biology contributes the predisposing condition but life experiences shape the resultant temperament.

Understanding the foundation for individual differences may enrich our scientific knowledge of brain and behavior but will it lead to meaningful applications of personality theory to real life problems and tasks? This has been a criticism of typologies and trait theories in general. Can one's biology be changed or, to paraphrase Freud, is biology destiny? It is conceivable that such knowledge could lead to improvements in drug therapies for mental disorders but could one's personality be enhanced by neurotransmitter manipulation? Could there be a time when there is a biological "cure" for excessive shyness, for example? Past trait and constitutional theories were tainted when they were applied to the concepts of eugenics. These are all questions beyond the scope of this discussion because they touch upon societal values and ethical principles. However, the historical analysis gives us a better context in which to evaluate contemporary personality neuroscience.

11. Detoxing the Brain

TODAY WHEN ONE TALKS ABOUT DETOX, it typically concerns one of two things. Individuals with alcohol and other drug addictions "go to detox" or, from a nutritional perspective, they are referring to some sort of purge to rid themselves of unwanted toxins in the body. The former nowadays usually means undergoing medical supervision while one endures the adverse effects of withdrawal from the substance of dependence. This detoxification process lasts until the remnants of the drug or drugs are no longer in the body. The latter is often an alimentary purge, either oral or in the form of enemas but sometimes achieved with special diets or dietary supplements. The focus of this chapter is centered mostly on drug detox efforts, while the chapter on Gut and Brain touches on the role of diet in detoxification. Historically, however, there has sometimes been little distinction between the two meanings and so there inevitably will be some overlap of discussion.

A major difference between contemporary and historical drug detoxification is that historical methods sought to cure the addiction whereas modern approaches tend to address only the withdrawal phase of rehabilitation. This difference is due to the relative understanding then and now of the addiction processes in the brain and even the differing definitions of what constitutes "addiction." The conflation of alcohol as food, intoxicant, and medicine also resulted in confusion in medical discourse concerning its use and abuse. The conceptual differences were not only influenced by scientific and medical developments but were themselves culture-bound.[1] For example, prior to the widespread use of distilled spirits, alcohol was an everyday beverage in Europe and America; indeed, for many individuals, alcohol was safer to drink than the available water.[2] Furthermore, alcohol had value as a medication, viewed as a stimulant in small quantities. An interesting medical debate over the use of alcohol in treating nervous disorders occurred during the mid–1800s among mental asylum alienists, only to be revived during the American Prohibition era.[3] The vice of drunkenness became more associated with distilled spirits among the lower classes or undesirables of society. Beer, for example, was viewed as wholesome and nutritious whereas gin led to debauchery, family discord, and ruin. While George Cheyne, author of the influential 1733 book *The English Malady*, famously advocated eschewing fermented beverages and meat for a diet of milk and vegetables, that advice did not go over too well by the public.[4] Beer even retained its reputation for therapeutic value as recently as the American Prohibition era and it was specifically proposed to be an exception to the limitations of prescribing alcohol for medicinal purposes. Conversely, the dangers of dependency upon opiates were not unnoticed by physicians but thought to be more of an inherent problem of individual constitution among the lower classes. Some of the earliest writings about opiate use, such as Jones' 1700 book *Mysteries of Opium Reveal'd*, warn physicians about the adverse consequences of opium withdrawal.[5]

In the 18th and 19th centuries, physicians distinguished between the vice of over-indulging in alcohol and the disease state of inebriety. One could be a drunkard but not an alcoholic.[6] The distinction was typically within the provenance of the patient and not the physician.[7] Inebriety was often seen as the larger disease category of which alcohol abuse was a subset. "Inebriety was a disease of 'civilized men' unable to bear the effects of alcohol and other types of drugs which had long been utilized by others."[8] However, the terms were sometimes interchangeable and reflected the shift, particularly in the latter decades of the 19th century, from treating addiction as a simply a consequence of immoral character or weak will to a brain disease model. The theory of causation dictated the treatment. Inebriety was sometimes viewed as incurable as it related to the nervous constitution of the individual, yet an individual could potentially achieve sobriety.[9] When addiction began to be viewed as a nervous disorder with a physiological cause within the brain, the treatments likewise became more medically oriented. Addiction was treated the same as any number of functional nervous system disorders. Like so many nervous diseases, there might not be an organic lesion but clearly there was an abnormality in the functioning of the nervous system.[10] This change in the conception of addiction was concomitant with the ascendancy of neurology as a distinct medical specialization.

German physician Eduard Levinstein was influential in his promotion during the 1870s of addiction as a neurological disease.[11] Rather than focusing on the will power of the patient, Levinstein believed that the physiological action of the opiates themselves could result in dependency. He did acknowledge the psychological changes wrought by chronic use of opiate and particularly the adverse effects on character but thought those changes were an effect of the drug rather than an inherent constitutional flaw. Long-term opiate use inevitably resulted in a changed personality—and not for the better!

In the decades spanning the mid–1800s to 1920s, treatment for addiction focused predominantly on withdrawal from the abused substance and in those days, successful withdrawal constituted cure.[12] Because of the emphasis on addiction as a disease, treatments utilized available medicines to manage the adverse effects of withdrawal. However, in some instances, the drugs given aimed to eliminate the toxins that were presumably accumulating from the addictive drug. For example, the "autotoxin" theory, emerging from new developments in immunology and bacteriology, stated that opiates produced toxins in thc gastro-intestinal tract. Purging the gut was therefore a mandatory feature of treating opiate addiction.[13]

Specific remedies for alcohol, tobacco and opiate addiction were usually indistinguishable by ingredients, if not by marketing, in the 19th century. Applications of addiction treatment by drug remedy had generally taken the form of one of two approaches.[14] "Cold turkey" with abrupt cessation of the substance characterized the Levinstein treatment.[15] Withdrawal was accomplished under the supervision of a physician in the confines of a locked and secure room and supported with various medicines ranging from belladonna to strychnine to cocaine.[16] Strychnine and atropine were common ingredients in the patent medicines sold for addiction therapy. These substances typically acted as drug antagonists and ameliorated the adverse intestinal effects of withdrawal. Cocaine was also believed to ease withdrawal effects but unknowingly, physicians were contributing to cross-tolerance between the drugs. Bromides, hyoscine, heroin, cannabis, and early barbiturates were given for sedation.[17] Electric stimulation, hydrotherapy, hypnotism, bleeding, and other somatic treatments were sometimes applied as adjunctive therapies.[18]

The second approach in the medical treatment of addiction was to enlist the family, usually the spouse, to give the anti-drug remedy surreptitiously.[19] Again, these remedies tended to be stimulants and/or emetics with names like the White Star Secret Liquor Cures (sold in the Sears Roebuck catalog) or the Dr. Haines Golden Treatment (sold by the Golden Specific Company)—odorless and tasteless so it could be mixed in beverages or food. Dipsocure was another purported cure for alcoholism that could be covertly administered "in the hands of a devoted woman."[20] It not only de-activated the alcohol, the remedy restored the nerves. Importantly, it also promised to make alcoholic men avoid alcohol entirely. Chemical analysis revealed that Dipsocure was composed of a sedative, potassium bromide and an analgesic relative of acetaminophen, acetanilide. These were drugs that promised to cure alcoholics of their drink cravings against their will and without them even knowing they were being treated!

Some of the time the ingredients were innocuous and contained little more than cornstarch and glycerin. More insidious were the remedies for addiction that contained the very substances that they were meant to combat. The Teetolia treatment advertised cures for alcoholism in only four days with its undisclosed combination of 30 percent alcohol and quinine.[21] The Richie Morphin Cure, advertised as a mail-order cure for morphinism contained 2.12 grains of morphine per dose![22] One could argue that this approach was actually analogous to the present day practice of maintenance therapy with methadone in that dosages were prescribed and there was less chance of graduating to more dangerous substances or using unhygienic methods of administration. Prestigious medical journals reported at length chemical analyses of various patent medicines and over-the-counter nostrums in order to educate physicians who were often treating the patients who had used those remedies but had failed to be cured by these quack remedies.[23]

The Keeley Cure, a popular American 19th-century treatment for alcohol and drug abuse, was one of many addiction cures during this age of patent medicines.[24] Others included Ensor Vegetable Cures, Dr. Meeker's Antidote, and Coca-Bola.[25] The Keeley system was created by Dr. Leslie Keeley (1834–1900) in the 1870s and remained popular for several decades. Inebriety, both in the form of alcoholism and opiate drug addiction, was a significant social problem in both the United States and Britain. The great irony of the proprietary addiction cure medicines was that the ingredients often consisted of alcohol and/or opiates in addition to a variety of purgatives and emetics. Thus, a long-term customer would be secured, content in the belief that the cure was keeping the addiction at bay.[26]

Keeley developed a comprehensive therapy that utilized features addressing the physical, psychological and social elements of addiction. He had a residential treatment program characterized by several weeks of intensive therapy guided by principles that in later years would be adopted in the creation of Alcoholics Anonymous.[27] A civil war surgeon, he initially had developed a treatment program for alcoholism in a Union hospital during the war.[28] After the war, he worked with Frederick Hargreaves, a minister who gave lectures on the temperance circuit, to hone his treatment program. This would develop into the Keeley Cure. "Drunkenness is a disease and I can cure it," he proclaimed in subsequent tracts and advertisements.[29] Furthermore, in contrast to some of the degeneracy theories of alcoholism that were prevalent in the late 19th century, he believed that inheritance had only a limited role in disease and that children of alcoholics developed their own drinking habits via observation.[30]

The first Keeley Institute for the Treatment of Inebriates, basically a sanatorium, opened its doors in 1879 in Dwight, Illinois, Dr. Keeley's hometown. Treatments there featured both individual and group counseling, physical exercise, hydrotherapy, lectures, and prayer meetings.[31] Moral support from one's peers was an important aspect and a group of residents would be on hand to greet new arrivals at the train station, escorting them to the Institute. Under continual supervision by an "attendant" who was assigned to each patient, there was a "camaraderie with like-minded patients brought together for several weeks of intensive therapy."[32] Unlike typical sanatoriums and hospitals for inebriates, patients were unrestrained and free to come and go at will. The only requirement was to queue at the appointed hours for their "bichloride of gold" injections.[33] While the majority of Institute patients were males, local boarding houses ("Ladies Home") provided accommodation for female patients. Tracy suggests that Keeley's success lay in "replacing the manly camaraderie of the barroom with that of treatment."[34]

The detox portion usually lasted nine days until the worst of withdrawal symptoms had passed.[35] The complete "cure" was typically four to eight weeks, depending on the substance abused and other comorbid afflictions, such as neurasthenia. The treatment for opiate addiction was the longest duration. A typical fee in the late 1890s was $25 per week plus $6–15 per week room and board, and $3 per day for the attendant. One estimate is that more than 14,000 individuals took the Keeley Cure at one of the many institutes nationwide in just a one year period between 1892 and 1893.[36] Another estimate by a contemporary observer indicated there would be 600 to 1000 men at any given time just at the one Institute in Dwight.[37] Some states enacted "Keeley laws" to subsidize treatment for the indigent, such was its national reputation.[38]

Franchised Institutes sprang up in four cities in 1891, spreading to over a hundred across five countries only two years later[39] There was a buy-in fee of up to $50,000 and franchise owners then paid a percentage of patient fees directly to Keeley. Staff received their training, as well as purchased all the medicine, at Keeley's institute in Dwight. Revenues topped $2.7 million in the first half of the decade. Keeley received promotional boosts from testimonial books written by recovered addicts who had been treated at the Keeley Institute. He gained national prominence from a direct challenge issued to the publishers of the *Chicago Tribune*: Keeley said that he could cure any six alcoholics sent to the Institute by the newspaper in four weeks. The newspaper accepted the challenge and Keeley was true to his word.[40] The resulting publicity was all that an entrepreneur could hope. Keeley also obtained government contracts to treat addicts and alcoholics in several veterans' homes, including more than 1500 veterans in the Fort Leavenworth home.

While neurasthenic patients might be administered "Kola" (most probably a cocaine tonic) to strengthen the nerves, the cornerstone of the Keeley Cure for addiction was billed as "bichloride of gold" or "double chloride" of gold.[41] The gold cure was administered to the residents four times per day in ritualistic fashion.[42] The remedy was packaged specific to the substance of abuse and priced differentially, presumably containing slightly different formulas. They were sold in pairs in distinctly shaped and colored bottles to distinguish the targeted user. Tobacco users paid the least ($5 for the pair), followed by neurasthenics ($8), alcoholics ($9), and, lastly, the opiate users ($10). The remedies were also available to nonresidents by mail order from the Institute. Dormandy notes that Keeley copyrighted the bottle design and labels yet never patented the remedy itself.[43] To do so would have meant revealing the gold cure's ingredients and that was avoided at all costs.

Even today, the exact formula of the gold cure has never been completely understood.[44] Staff at the Keeley Institutes signed nondisclosure agreements as a condition of employment and chemical analyses published in medical journals reported inconsistent findings. Concoctions consisting of alcohol, opium, morphine, coca, strychnine (a stimulant), apomorphine, aloes, willow bark, ginger, ammonia, belladonna, atropine, hyoscine, and scopolamine were all variously reported.[45] Bi-chloride of gold not only contained no gold but consisted of some of the very ingredients that sustained the drug addiction. The regular injections four times a day can now be understood as a maintenance program to avoid withdrawal. Certainly the "cure" ensured a loyal clientele both from the Institute's mail order customers and former residential patients.

The remedy was not presented as a maintenance drug, of course, but rather a detoxification agent. Keeley wrote that chronic opiate use "created an 'isomeric change in the structure of the nerve and its action,' which then required the continual presence of the drug for the normal functioning of the nervous system."[46] This was nonsense even at the time, according to accepted physiological principles. However, in the latter decades of the 19th century, there was a concern among the medical community about the actions of poisons and toxins in the body. If not for the fact that there was no such thing as bi-chloride of gold, Keeley may have sincerely believed it when he said that the medicine worked by "speeding up the restoration of poisoned cells to their pre-poisoned condition."[47] Likewise, he attributed alcoholism to a poisoning of nerve cells.[48] "In poisoning, the battle is between the nerve cells on one side and the poisoning power of alcohol on the other. The cells resist the poison.... The poisoned nerve cells demand the presence of alcohol in order to subserve their perverted functions."[49] The "perversion" refers to the atavistic effects of alcoholism, a common belief during the Victorian era that addiction led the drinker to regress to a more primitive state of existence. However, Keeley maintained that the atavism was in the nerve cells themselves.

Keeley claimed an astonishing 96 percent cure rate.[50] The Keeley Cure was a national household phrase and well-integrated into popular culture: songs, political posters (notably one for Grover Cleveland's presidential campaign), and cartoons. Published testimonials were in abundance and Keeley boasted that the Cure had been adopted by the U.S. government to treat addiction in National & State Soldiers' and Sailors' Homes.[51] Keeley's patients were represented among 55 different occupations, including judges, physicians, and congressmen, although most were farmers.

One unusual feature of Keeley's operation was that he was willing to employ physicians who were recovering addicts themselves.[52] Personnel record at his institutes indicated that approximately a quarter of the 418 staff physicians had a history of addiction. The physicians supervised the acute detoxification stage of patients, administered the injections of the bi-chloride of gold, and provided medical lectures that constituted patient education for the residents.[53] However, by the early 1900s, the Keeley Institutes no longer employed physicians with past addiction problems, perhaps because of issues related to relapse on the part of the physicians, incompetence, and increasing public criticism of the practice by the medical establishment. Dr. T. D. Crothers, an asylum superintendent, believed that Keeley Institute practice of hiring went against professionalism in having an "unreliable contingent of ex-drunkards in the discipline."[54] A notice of the 1892 annual meeting of the Hampden District Medical Society in Massachusetts included a resolution to censor one of its physicians who had taken employment at a Keeley Institute.[55] It is noteworthy, however, that today many drug and alcohol rehabilitation centers rely on counseling staff who are themselves recovered addicts. While Keeley's motivation may have been based more on finding an employment pool of willing and licensed physicians who otherwise would have been loath to work in a fringe operation that was considered quackery by mainstream medicine, today's practice is centered on the empathetic authority and shared experience that a recovered addict can offer to those undergoing treatment.

The other key staff member in these institutes was the personal attendant, often an individual who was also a recovered addict.[56] These attendants were assigned one-on-one to patients at the beginning (and the most difficult part) of treatment. As mentioned previously, the Institutes were different from the hospitals and asylums in that patients were completely unrestrained. There were, however, a set of rules given upon admission that included the receipt of four daily injections of whichever version of the gold cure was suitable, attendance at the medical lectures, abstinence from smoking, gambling, or sodas (considered tonics back then; coffee and tea were permitted but only in small quantities), riding in cars, or interacting with patients of the opposite sex.[57] The most curious feature of the program, perhaps, was the allowance of whisky as desired by the patient. Patients were provided all the whisky they wanted but were not allowed to bring any in from the outside. This practice presumably warded off the delirium tremens and seizures associated with alcohol withdrawal and supposedly they "lost the appetite for it—usually within three or four days."[58] Of course, this reduction might be attributed both to the alcohol contained in the remedy and the aversive conditioning to some of the ingredients (e.g., apomorphine) and thus patients were able to affect a maintenance level. Opiate addicts were weaned gradually from their drug.

With the daily injections and accompanying tonics, patients improved in the supportive environment where there was plenty of rest, nutrition, social support, and healthy exercise.[59] Recovered patients were told upon departure that they were cured unless they relapsed, in which case their nerve cells would be poisoned again. They were given discharge orders that included the maintenance of many of these features: regular sleep, meals, plenty of water, recreation, and "care in the selection of personal associates."[60] They were encouraged to write weekly to the Institute, fellow patients (either still in residence or recovered), and to continue a regular maintenance dose of the gold cure. Recovered addicts formed alumni chapters, called the Keeley League, at one time numbering 30,000 among

340 chapters nationwide. The goals of these clubs were prevention of intemperance and maintenance of sobriety and to promote the Keeley cure. Their motto was "We were once as you are; come with us and be cured."[61] The clubs not only provided social support in ways similar to Alcoholics Anonymous today but had an activist role to encourage the passing of Keeley laws. There were even clubs for family members of patients, much like Al-Anon family support groups.

The purported success of the Keeley cure did not go unnoticed or unchallenged. Keeley's stated 95 percent cure rate (even higher for opiate addictions) did not include patients who left treatment prior to completion of the program. In fact, the reporters from the journal *Christian Advocate* surveyed former Keeley Institute patients and reported a 49 percent recidivism. Criticisms focused on his claim of primacy in viewing addiction as a brain disease when asylum alienists and neurologists had been saying that for over a decade.[62] The British medical community engaged in spirited debate in 1892 on the merits of "the double chloride of gold cure" and whether Keeley's cure should be officially sanctioned.[63] The Society for the Study of Inebriety commissioned a chemical analysis, which revealed that the remedy contained various "poisonous intoxicants" as well as 27 percent alcohol.[64] The franchising operation was denounced as "a commercial company for pushing the golden remedy," advertising that "without any effort on the patients' part a cure would be effected by the use of this wonderful and infallible remedy."[65] Furthermore, a representative of the Society had visited the Keeley Institute in Illinois and reported that Keeley refused to reveal details about the treatment program. The result of the Society meeting was a motion that secret cures for inebriety could not be recommended by the Society or the medical profession in general.

The medical community overall viewed the gold cure as quackery, the business as crass commercialism, the advertising unprofessional, and Keeley's refusal to disclose the ingredients was considered a breach of medical ethics.[66] Besides fear of competitors, Keeley maintained that his cure consisted of more than the medicine itself but instead, it was a whole system and thus it could not be scientifically replicated by others.[67] He also noted that many of his critics were alienists or asylum doctors and consequently, as competitors, they were inherently biased against his system. The chemical analyses that were performed in the mid–1890s had revealed that not only was there no gold in the remedy but instead there was a variety of potentially harmful substances that could produce serious side effects. How, Dr. T. D. Crothers, an alienist, wondered, could doctors treat the side effects if they didn't even know what was in the drug?[68] If the remedy was genuine, then Keeley should have nothing to fear and should, in fact, welcome validation by the scientific community.[69]

The Keeley Institutes' popularity peaked in the 1890s, although they hung on throughout the first half of the 20th century.[70] A stream of lawsuits and countersuits brought unwanted publicity and by the turn of the century, most of the institutes had closed. Most damning perhaps was the 1902 court testimony in one of the lawsuit trials by Keeley's original collaborator, Hargreaves. His business interest had originally been bought out by Keeley in 1886 and he subsequently became one of Keeley's competitors. According to his sworn testimony, the gold cure had only contained gold at the beginning of their partnership and was hastily removed after their first patient died from taking it. Afterwards, gold was added to a few samples just for analysis and not for patients. Furthermore, the early patient testimonials were written by Keeley and himself and not real patients. The cure was a placebo, according to Hargreaves, and the medical journal articles a fabrication of evidence by Keeley. In 1925, the gold cure was discontinued in favor of the other tonics manufactured by the Keeley Institute. By this time, Leslie Keeley had been dead for a quarter of a century (having died a millionaire) and the business was owned and operated by the family of the chemist who was one of Keeley's original partners.[71]

However, the most likely contributor for the abrupt decline of the Keeley cure was the legislative passage of the Eighteenth Amendment to the U.S. Constitution in 1919 that mandated the era of Prohibition.[72] By 1935, there were only four Keeley Institutes remaining in the United States. The last and original one in Dwight, Illinois, finally closed its doors in 1966.

As White points out, while the gold cure itself was quack medicine, the overall scheme introduced principles of rehabilitation that are still in widespread use today.[73] Some of those principles are clearly evident in the precepts of Alcoholics Anonymous, for example. The notion of addiction as a brain disease did not originate with Leslie Keeley despite his claim to the contrary but Keeley's promotion of that view nevertheless permeated into the popular culture pre–Prohibition.[74] This helped the shift, also being advocated from many physicians and asylum alienists, from the moral model that had dominated during most of the Victorian era.

The conflict between the moral and disease models of addiction were evident in the medical establishment's reaction to the Keeley cure. Keeley couched his remedy in a "single cause/single cure" paradigm, much as others had done in promoting the new germ or microbe theory of disease.[75]

While undoubtedly the purgatives and emetics contained in his remedy were manifestly an aversive conditioning treatment, this medicinal approach was not unique to Keeley's cure.[76] His use of social support and individualized attention was unique, however. Furthermore, he was able to blend the conception of alcoholism as a brain disease with the popular degeneracy theories of the day. While he did not believe that alcoholism was itself a heritable trait, one could inherit a "weakness of resistance" to the drink—in other words, a diasthesis model of addiction.[77] Contemporary theories of alcohol addiction also speak to the genetic predisposition possessed by some who abuse substances.[78]

The search for an addiction cure by modern neuroscience more than a century later has not yielded the results hoped for by advocates of the disease model. Just as in earlier times, drugs have been used to ease the adverse symptoms associated with withdrawal from the substance of abuse. Indeed, this process exemplifies the current meaning of the term "detoxification." The procedures are deployed with medical supervision and have reasonably standardized protocols. The most significant recent debate over protocols concerned the emergence of so-called "rapid detox" models.[79] Instead of the typical three to 15 days for detoxification, rapid detox aimed to produce the same results in just one to two days. More radically, "ultra-rapid" detox achieved cleansing in just a few hours. In both methods, the patient is put under anesthesia so that the intense discomfort experienced during withdrawal is presumably unperceived by the patient. On the surface, the advantages of a more rapid detoxification seem self-evident: patients experience less distress, are in the medical setting for considerably less time, and medical costs are reduced. Unfortunately for addiction medicine, there was both an absence of empirical data indicating better long-term outcomes compared to traditional protocols and an unacceptably high mortality rate associated with the procedure. The latter issue prompted the Centers for Disease Control and Prevention to issue a warning in 2013 against the procedure.[80]

Effective drug therapies for curing drug addiction have remained elusive and, according to Helfand, all existing attempts are performed in conjunction with counseling.[81] Some treatment drugs, like Antabuse, interfere with the normal metabolism of alcohol, resulting in aversive effects, such as severe nausea, when taken with the abused substance. This is a form of counterconditioning: learning to associate the substance of abuse with negative physiological symptoms, such as nausea and vomiting. The addicted individual then tends to avoid the substance. In the 1950s Antabuse (disulfiram) was the first drug to treat alcoholism approved by the FDA and still used in some treatment facilities today. Antabuse was heavily touted as an adjunct therapy for alcoholism but has had only modest results mostly because of compliance issues. Additionally, because it acts upon alcohol in any form, there can be complications from interacting with other medicines or common products that contain even small amounts of alcohol (e.g., using isopropyl alcohol to disinfect an area on the skin or applying many skin cleansers).

Other addiction medications have been utilized under the guise of substituting a less harmful drug for another. Methadone is a good example here. Methadone, while also an opiate, is longer acting than heroin and many other drugs in that class and does not produce the same level of euphoria. This is used as a maintenance drug to stave off withdrawal symptoms and administered in outpatient clinics according to appropriate hygienic standards. This type of drug therapy is supportive or maintenance therapy and not a cure.

The third category of addiction drug therapies are neurotransmitter antagonists. Naloxone (Narcan) and naltrexone are in this category and feature much in the news these days as drugs that can reverse an opiate overdose within minutes, even seconds in some cases. The drugs block the receptor sites that opiates bind to and thus render the opiate ineffective. However, again, this approach is not curative and does nothing about future cravings. Instead they are virtually an instantaneous provocateur of withdrawal. Current neuroscience research on drug addiction suggests that there is a distinction between "liking" and "wanting" and the neural pathways that underlie each of those states are distinct and are mediated by different neurotransmitters.[82] Thus, drug therapies will have to become more targeted.

Have we learned our lesson from the days of quack cures with their promises of quick, painless relief from drug and alcohol addictions? Unfortunately, there are any number of so-called "cures," home remedies, and non-regulated dietary supplements touted on the Internet for the treatment of addiction or withdrawal symptoms. But is the conventional medical community immune? PROMETA entered the addiction therapy scene a few years ago as a much heralded miracle cure for addiction to methamphetamine. The drug was actually a new combination of three existing, FDA-approved medications. A curious loophole in the FDA regulatory system is that combinations of drugs already approved singly can be marketed without additional study. Furthermore, the combination drug can be targeted for a condition never tested in the studies done on the individual drugs.[83]

The PROMETA combination consisted of gabapentin, flumazenil, and hydroxyzine. All three drugs affect various neurotransmitters but their specificity for countering methamphetamine addiction is questionable. Gabapentin is a GABA agonist, an anticonvulsant with analgesic and anxiolytic properties. A few research studies have indicated reduction in addiction-related cravings for cocaine.[84] Flumazenil also affects GABA and is typically used to treat overdoses of benzodiazepines as a specific benzodiazepine antagonist. According to Ling et al., this is purported to be the key ingredient in the PROMETA formulation.[85] The third drug in the combination is hydroxyzine, a sedative often used during detoxification procedures to help alleviate anxiety and nausea. In contrast to the GABA-ergic properties of the other two drugs, hydroxyzine is basically an antihistamine. While the various physiological effects and mechanisms of these three drugs are all potentially relevant to detoxification, there is nothing theoretically to suggest a particular affinity for curing methamphetamine addiction. As Ling et al. point out, the action of methamphetamine is most notable on dopamine pathways yet these are not the targets for the three individual drugs composing PROMETA.

Initially, the creation of PROMETA and its subsequent appearance on the medical scene did not appear to be outside the norms of pharmaceutical research and development despite the founder of the company being a junk-bond salesman with no medical or scientific credentials.[86] The telltale signs of quackery began to appear, however, with the marketing. In the absence of rigorous, controlled research studies, testimonials and anecdotal evidence of almost perfect cure rates (98 percent) were advertised. The cure rate, as defined by abstinence, was far above typical drug rehabilitation programs. A massive media marketing campaign with endorsements from experts resulted in patients flocking to treatment centers. These centers were franchised, for-profit clinics that charged as much as $15,000 per treatment.[87]

Problems for PROMETA began when journalists and the medical community began a closer scrutiny of the operation.[88] Treatment results had been grossly exaggerated and samples manipulated to exclude patients who had dropped out of treatment. Government officials who had helped secure taxpayer-funded grants to do these initial studies showing such spectacular results were found to have significant financial interests in the company. Finally, results from a placebo-controlled, double-blind study (the gold standard design for pharmaceutical research) revealed that the PROMETA therapy was no more effective than a placebo.[89] And so another miracle cure for addiction was debunked!

Another supposed cure for addiction that has made some headlines is the hallucinogen, ibogaine. This is a drug compound synthesized from a plant, *Tabernanthe iboga*, found in the tropical forests of Mexico and Central America and is illegal in the United States. However, proponents maintain that the drug is a cure for addiction, reporting 70 percent success rates.[90] Like with PROMETA, patients have been willing to pay large sums of money to go abroad for an untested and unregulated treatment that comes with huge promises. Closer inspection of abstinence rates show that the numbers are manipulated to exclude individuals who have not finished the treatment. More concerning is the unacceptably high morbidity and mortality results associated with ibogaine use—hence, its failure to achieve legal status in this country and the difficulties in conducting human research safely. Unlike PROMETA, ibogaine's pharmacological action is on the dopamine pathways involved in the brain's addiction pathway.[91] The hallucinogenic properties of the drug arise from serotonergic pathways, just as with other hallucinogenic drugs, such as mescaline and LSD. Those drugs were notably used in research on treating alcohol dependence in the mid–20th century with some success, as discussed in Chapter 6. While ibogaine treatments have been relegated as an unproven and potentially quack cure, limited scientific research has begun with some of its chemical derivatives. Only future scientific studies with proper controls will reveal whether this a finally a path to a chemical cure to addiction.

12. The "Rest Cure" Revisited

> "If a physician of high standing, and one's own husband, assures friends and relatives that there is really nothing the matter with one but temporary nervous depression—a slight hysterical tendency—what is one to do?"—CHARLOTTE PERKINS GILMAN, *The Yellow Wall-Paper*, 1892, 1

SILAS WEIR MITCHELL (1829–1914) has been known variously as a founder of American neurological practice, experimental physiologist, toxicologist, novelist, and, most often today, representative of the Victorian male medical hegemony responsible for oppressing women. His "Rest Cure," a treatment made famous in the 1892 short story by Charlotte Perkins Gilman, *The Yellow Wall-Paper*, came to symbolize the suffocating male dominance in the late Victorian era.[1] While no longer practiced today, the Rest Cure was a systematic therapy for neurasthenia, an ill-defined nervous disorder diagnosed mostly in women in almost epidemic proportions in the last decades of the 19th century. This chapter will revisit the rest cure in the context of the contemporary psychological and physiological theories of its day, examine the practice as described in *The Yellow Wall-Paper*, and finally, offer some alternative interpretations of the treatment.

We apparently live in the "age of anxiety." At least, headlines from the news media tell us so: "'We live in an age of anxiety'—Surge in mental health problems in last six years"; "Campuses face mental health crisis"; "Why are more American teenagers than ever suffering from severe anxiety?"[2] However, a recent epidemiological analysis questions this assertion.[3]

Along with assessing in 2015 the prevalence of anxiety in the population with a cross-cultural comparison, researchers examined the evidence to answer the specific question of whether the prevalence had indeed increased, as suggested by media reports.[4] They noted the difficulties in long time-span analyses due to changes in diagnoses, the source of the data (e.g., whether from community surveys or clinical surveys—meaning self-report versus professional assessment), and changes or differences in cultural attitudes that influence disclosure and seeking treatment. The analysis showed that approximately 30 percent of Americans experience clinical anxiety with the average case being a middle-aged female (although they note that the median age of onset is 11 years). Related to the question of whether there has been a rise in clinical anxiety disorders, the data did not show support for that hypothesis. There has been virtually no change since 1990 when epidemiological studies began to have consistent methodologies that allow for longitudinal analysis. Genetic studies have also shown a consistent heritability factor, ranging from 30 to 50 percent.[5] What has perhaps changed in that quarter century is an increase in the rate of those seeking treatment.[6]

There have been other time periods in our history which have been billed "an age of anxiety." *Fortune* magazine in the 1930s estimated that there were millions of neurotics in the United States.[7] Examination of pharmaceutical prescriptions reveal that the barbiturates, such as the best known phenobarbital, were being prescribed by the millions during World War II. A 1942 issue of *Newsweek* suggested that seven million (or one in seven) Americans suffered from anxiety disorders.[8] Barbiturates were the first class of minor tranquilizers used primarily to treat anxiety and insomnia in the first half of the 20th century and were very popular during the 1920s and '30s. Barbital, later renamed Veronal, was the first barbiturate to be marketed in 1903 and gradually replaced the various bromides (e.g., potassium bromide) that had long been in use during the 19th century as a sedative. Because of the heightened overdose potential, especially when used in conjunction with other nervous system depressants, such as alcohol, barbiturates are now rarely prescribed for self-administration. Some barbiturates are currently prescribed to prevent epileptic seizures or for their anesthetic properties.

The advent of minor tranquilizers and antidepressants in the 1950s was aimed at treating masses of anxious and depressed Americans who could not afford psychotherapy. Prescriptions of "happy pills," such as Miltown in '50s and '60s and Valium in the 1970s became widespread among ordinary Americans.[9] Protected by the cloak of medical respectability since these were all prescription medications, use of these drugs increased dramatically. The National Institutes of Health reported in the late 1960s that over half of American adults had tried prescription mood-altering drugs.[10] Prescriptions of Valium only decreased in the 1980s when newer minor tranquilizers and antidepressants, such as Xanax and Prozac came on the market and replaced it as the leading psychoactive prescriptions.

But was there truly an epidemic of anxiety fueling the dramatic rise in use of these drugs? Numbers do not support that popular media conclusion. We can more easily account for the increase by the changes in regulations regarding pharmaceutical advertising and the rise of the patient's rights movement.[11] Direct-to-consumer advertising not only put information in front of patients so they might be more inclined to ask their physicians about the drugs, it perhaps gave the impression that usage was widespread and normative.

In the 1880s there was an epidemic of functional nervous disease sweeping across the United States. Neurasthenia, or "nervous exhaustion," led the pack of disorders and for a time, it was a common, if not the most common of the diagnoses. George Beard initially described the disorder in 1869, suggesting that it arose from the stress of post–civil war and the concurrent rise in industrialization.[12] It was characterized by a wide array of symptoms ranging from chronic fatigue to depression to dyspepsia (i.e., disturbances of the gastro-intestinal tract). The disease was especially associated with the unique features of American society, so much so that a slang term for the disorder, "Americanitis," was coined.[13] Neurasthenia was a "growing pain" of the modern society.[14] The disease was also correlated with socio-economic status, education, and sensitivity because those individuals were thought to have more vulnerable nervous systems.[15] Once accepted by both patients and physicians alike as a diagnosis for the panoply of vague symptoms, its manifestations were seen everywhere.[16]

The symptoms were vague and ill-defined: "inability to sleep, loss of appetite, listlessness, and a variety of aches and pains."[17] Physicians noted anxiety, hypervigilance and agitation comingled with melancholia and hysteria, although neurasthenia was soon differentiated from the latter two disorders. In all, Beard cataloged three dozen symptoms.[18] It became a "catchall diagnosis applied to a syndrome with vague and variable symptoms ... a variety of nonverifiable physical complaints" that had no apparent organic cause.[19] In the absence of clear understanding the physiology of nerve function, medical explanations were equally vague but metaphors regarding spent energy and depleted bank stores were common. Beard's contribution was that he provided a theory to place neurological order upon the seeming disarray of symptoms. "Neurasthenia is a chronic, functional disease of the nervous system, the basis of which is impoverishment of nerve force, waste of nerve-tissue in excess of repair; hence the lack of inhibitory or controlling power—physical and mental—the feebleness and *instability* of nerve action, and the excessive sensitiveness and irritability, local and general, direct and reflex.... Nervousness is really *nervelessness.*"[20] The fatigue and agitation were clearly malfunctions of the voluntary muscles, the nerve cells themselves must be fatigued and thus be treated with stimulants, and increasing emotionality must be associated with increasing nervous exhaustion along a continuum of nervous complaints.[21]

Even though Beard categorized neurasthenia as a functional nervous disorder, that only meant that an obvious organic nervous system abnormality was not apparent according the known neuroanatomy of the day. He still attributed the vast array of symptoms to a single cause, however, and believed that in the future with new technology (either through chemical analysis or microscopy), structural abnormalities would be detected.[22] Beard's depiction of nervous exhaustion clearly built upon his contemporaneous theories of nervous system action: reflex theory (as espoused by Pavlov and Marshall Hall), the electrical nature of the nervous impulse (Du Bois-Reymond's law of excitation), heritability of acquired characteristics (Brown-Séquard), and closed system thermodynamics (Helmholtz and Meyer's theories). No two individuals had the same nervous energy and likewise, no two individuals had the same environment. Neurasthenia was the result of an excess demand of nervous force due to environmental demands. The bodily centers where nervous energy was most concentrated were the brain, the gastro-intestinal tract, and the reproductive system. Thus, there was a physiological link with dyspepsia (and other gut complaints) and hysteria or excessive masturbation. Both over-use and under-use could lead to problems because of the reverberations in the nervous system from reflex action.[23]

Initially, neurasthenia was diagnosed predominantly among males, particularly the "brain workers."[24] William James was a famous example and suffered from neurasthenia off and on throughout his life. Despite the modern depiction of Theodore Roosevelt as the epitome of the energetic outdoorsman, he, too, was a famous neurasthenic patient. But for the American neurologist Silas Weir Mitchell, the typical case was a young female, 20–30 years old, and often having experienced the stress of nursing a sick relative.[25] They also tended to be thin, anorexic, and be surrounded by over-attentive female family members. By the time Mitchell was consulted, the patient often would have already tried a succession of home remedies, tonics, patent medicines, and treatments prescribed by the family doctor. Significantly, comorbidity with drug dependence was observed, although the term not in use then.

Not surprisingly, treatments ran the gamut of mainstream medicine to charlatan quackery. Cocke in 1894 provides an example case study that is fairly typical and expresses the vast array of treatments applied:

> First, bromide of soda and chloral, morphine in combination with atropine, phenacetine with and without bromides, antipyrine, chloro-dyne, whisky, brandy, cannabis Indica, sul-phonal, several of the valerianates, a number of the homeopathic potences of caffia. She had tried massage and warm baths, all for insomnia, in a period of four years.
>
> For her other symptoms she had attended most of the mineral springs of Europe. Was treated with the water-cure ... douched with cold water and with warm water. She had tried the Scotch douche. She had tried shower-baths. She had tried electricity with and without baths. She had consulted a number of specialists, and was treated by them for uterine and ovarian difficulties. She had tried lavage (washing out of the stomach). She had taken no less than eighteen of the salts of iron, in combination with phosphorus, quinine, cinchonidine, and cinchona. The salts of zinc had also tried their hand upon her. Specialists of the throat and nose had done their best for her. As she had had an occasional attack of hives (urticaria) the dermatologists had also done what they could for her. Her friends prayed for her, and she prayed for herself.... Magnetic physicians laid their hands upon her, but, alas! with no success.[26]

The emerging fascination with electricity let to a variety of electrical therapies to increase vital energy and restore the nerve resources. For men, electrical stimulation of the genitals was thought to be beneficial to counter symptoms of fatigue. Hydrotherapies, from hot baths, saunas, and wraps, were standard treatments for women, again for stimulating the body. Patients were thought to benefit from tactile stimulation, either in the form of massage "rubbing" or various vibratory mechanisms. Not only would the voluntary muscles be activated, the pressure or vibrations would help wayward uteruses and other organs "jiggle" back to their proper place.

Nearly universal, however, was the application of patent medicines. In fact, an entire subgenre of nerve and brain tonics and elixirs was born. Nerve tonics, such as Coca Cola and Dr. Miles' Restorative Nervine, were ubiquitous in drug therapies aimed at alleviating fatigue and nervous exhaustion. Other drugs in the arsenal to treat neurasthenia included sedatives like bromides (usually lithium but sometimes potassium), phosphates, or increasingly, alcohol-laced substances, such as Lydia Pinkham's Vegetable Compound, thought to restore health and vigor. The latter, one of the most popular patent medicines ever produced, was 18 percent alcohol, equivalent to fortified wine. Opiates, in the form of morphine, laudanum (tincture of opium) or undisclosed in a variety of pills and potions elixirs were prescribed daily to deal with vague aches and pains and insomnia that plagued the neurasthenic. Chloral hydrate, a relatively new sedative, was extensively used by physicians to treat insomnia. In other words, the typical neurasthenic was taking a daily cocktail of addictive drugs. Is it any wonder, patients were resistant to treatment and suffered from a confusing and sometimes contradictory set of symptoms that defied medical explanation?

De Paula Ramos arrives at a similar conclusion in the most famous 19th-century case of hysteria, Anna O.[27] Examining drug therapy and dosages in medical records and letters, he makes a strong case that most of her symptoms were consistent with addiction to chloral hydrate and opiates and their respective withdrawal syndromes. This is significant because this is one of the most iconic case studies in psychiatric history as outlined by Freud in his *Studies on Hysteria*.[28] Her case, actually relayed secondhand to Freud by Josef Breuer, her physician, was the first to describe the "talking cure," which became the foundation of psychotherapy.

Interestingly, while the therapies described above were widely employed in the treatment of neurasthenic females, neurasthenic males underwent therapies with a very different approach. For them, exercise and change of scenery were often prescribed. For example, the recommendation for Theodore Roosevelt was to go to a dude ranch in South Dakota; William James famously went abroad to Europe following the birth of each of his children.[29]

Silas Weir Mitchell

Silas Weir Mitchell, or Weir Mitchell, as he was usually called, was a Philadelphia physician and experimentalist, the son of another eminent Philadelphia physician. After a post–medical school year in 1851 spent in Paris with one of the world's preeminent experimental physiologists, Charles Barnard, he returned to Philadelphia to join his father's practice. He performed pioneering studies in toxicology, researching the toxicology of rattlesnake venom. This work helped with the development of antidotes for snake bites and this achievement alone would have secured Mitchell fame in the medical community and history logs. However, it was his research and treatments of gunshot wounds and paralysis in soldiers during the Civil War as a field and hospital surgeon, that propelled him to further prominence in the medical field. His theory concerning the phantom limb phenomenon remained unchallenged for more than a century until recent times. Post-war, he continued his medical practice, specializing in the emerging fields of neurology and nervous diseases, and published over 100 books and journal articles on related topics. He was also a successful fiction writer, publishing numerous novels, short stories, and poetry. Perhaps his most famous story, published in *The Atlantic Monthly*, was *The Case of George Dedlow*, based on a real Civil War neurological patient case. Mitchell became the most well-known American neurologist of his time, as well as a well-known public celebrity. His friends and correspondents from all walks of life are a "who's who" of the gilded age, from Oliver Wendell Holmes to William Osler to Havelock Ellis to Henry James. He was a correspondent with many of the early experimental psychologists, most notably, William James, when psychology was a fledgling academic discipline in the United States.

Weir Mitchell was also the first scientist in the United States to experiment with mescaline, the psychoactive drug found in the buttons of the peyote cactus. One of his contributions to psychology was a comprehensive description of the hallucinogenic experience that remains unchallenged as an early foray into psychopharmacology.[30] He gave some peyote buttons to William James, knowing his interest in exploring altered states of consciousness. James was keen on experiencing the intense hallucinations that Mitchell had reported and later published. James recounts that he wasted two days being violently ill and did not experience any of the hallucinations that Mitchell had described.[31]

Mitchell's friendship and professional ties with William James would soon be strained when James recruited Mitchell to participate in research on the supernatural. James was interested in the spiritualism movement and particularly the work of Mrs. Mary Piper, a popular medium. At James' request, Mitchell attended a séance and was not impressed. When he expressed his skepticism to James ("inconceivable twaddle" was his exact phrase), his colleague was annoyed and the friendship waned soon thereafter.[32]

He used his early training in experimental physiology to explore many psychoactive substances besides mescaline, to include lithium bromide, woorara (curare), amyl nitrate (for epilepsy), morphine, and sassy-bark.[33] Lithium remains today a standard pharmaceutical treatment for bipolar depression.

The Rest Cure, however, is what Weir Mitchell is remembered for most, particularly outside the field of neurology. As the United States entered the Industrial Age in the decades following the Civil War, there was concern about the strain of modern living. In 1872 Mitchell published a best-selling book intended for the general public, *Wear and Tear, or Hints for the Overworked*.[34] The book touched on that uniquely American disorder of the time: neurasthenia. The book was wildly successful yet criticized by the medical community for being too popular! The first printing sold out in ten days and five editions were published in the first decade of its release. Ironically, it was precisely the publishing success and marketing to the general public that smacked of "unprofessionalism" to Mitchell's peers.[35] Nonetheless, the book did much to disseminate and popularize the notion of nervous exhaustion to the general public. Here he discussed the problems of the modern, industrialized frenetic pace and urbanization of American lifestyle, and the medical hazards posed to its citizens. Because heredity endowed one with only a fixed amount of nervous energy, when demand exceeded supply, nervous exhaustion or neurasthenia was the result.[36] "Every blow on the anvil [referring to muscular activity] is as distinctly an act of the nerve centres as are the highest mental processes."[37] Various conditions or attributes of the nervous system, according to Mitchell, contributed to the state of nervous exhaustion: first, the brain had no inherent warning system when its energy stores were being completed; second, urban crowding and extreme variations in climate conditions in the United States; and third, the poor health of American women, the source of which was their increased levels of education.[38] Dr. Mitchell was indeed a man of his time with respect to views towards women. Mitchell's evidence for the differential effects of education on women was that the number of female patients in their late teens were still in school at rates disproportionate to that of men.[39]

In *Fat and Blood,* Mitchell also makes the case for psychological illnesses having a biological basis and vice versa. "Diseases of mind could be cured by treating the body."[40] The latter point subjected Mitchell to professional criticism with accusations of confusing psychiatric and neurological disorders. The underlying principles of the Rest Cure are built upon a mixed foundation of psychological and neurological ideas, for which Mitchell was wholly unapologetic.

Mitchell's Rest Cure was an attempt to address the unique and frustrating aspects of the neurasthenic patient: the vague symptoms, anxiety, depression, anorexia, and fatigue. His 1877 monograph, *Fat and Blood,* outlined his "Rest Cure" for the medical community.[41] He describes some of the characteristics already mentioned in the typical patient, stating the treatment is mostly for "women of a class well known to every physician,—nervous women, who, as a rule, are thin and lack blood."[42] They need "constant stimulus and endless tonics. Then comes the mischievous role of bromides, opium, chloral, and brandy."[43] Overmedication of these women is mentioned more than once.

Interestingly, Mitchell offers "one fatal addition to the weight which tends to destroy women who suffer in the way I have described. It is the self-sacrificing love and over-careful sympathy of a mother, a sister, or some other devoted relative."[44] He talks much of the psychological interaction that the two have, with one creating a new victim in the other, an unhealthy dependency, and a craving for attention and sympathy by the patient. The illnesses in these patients were largely self-induced and enabled by those around them. In today's health psychology terminology, the patients utilized "secondary gains," taking advantage of their illness to escape household or familial responsibilities. "An hysterical girl is, as Wendell Holmes has said in his decisive phrase, a vampire who sucks the blood of the healthy people about her."[45] An even worse situation was when the patient's caregiver was of a nervous or irritable temperament: "Two such people produce endless mischief for each other."[46]

The Rest Cure was not universally applied in the treatment of neurasthenia but meant for the most serious cases. The condition typically had taken many months, even years, to become entrenched and so a quick fix was not likely. The rest period was meant for real rejuvenation and restoration of the nervous system. Absolute rest was required "for all the overstrained, contractured voluntary muscles gradually to let go; for the kinks and twists in the intestines and other internal organs to become un-kinked or unwound ... and for the agitated cells and nerves to come back to mobile equilibrium."[47] The enforced nature of the rest and seclusion also had a bit of reverse psychology attached: for the hypochondriac, the order for complete bed rest, combined with deliberate ignoring of complaints, produced compliance with the other measures, such as overfeeding to gain back weight, in an effort to rejoin family and company.[48] The treatment was also modified according to the patient's particular circumstances and needs, as well as adapted by the British to suit their cultural differences from Americans.[49]

The actual Rest Cure procedure, according to Mitchell was relatively straightforward.[50] First, the patient was removed from his or her habitual environment. This had two purposes: the seclusion provided a highly structured environment to render treatment and it also separated the patient from the enablers. Mitchell was adamant that a professional, emotionally-detached carer or nurse be employed and that the patient not be cared for by a relative or friend. This was in order to break the nonproductive, emotional enabling that went on in dealing with these patients. Second, strict bed rest for three weeks was prescribed. During this time, almost all medications were stopped and a diet implemented for weight gain and treating anemia. In fact, the first week was essentially a detoxification program and not so different from in-patient rehabilitation programs today. A difference, however, is that the Rest Cure did not need to be undertaken at an institution, such as an asylum or retreat, but rather could be achieved in a privately rented apartment or house with a hired nurse—just so long as it was a different locale from the home.

During the daily physician visits, there was to be no discussion of the patient's symptoms within the hearing of the patient; instead the patient was encouraged to talk about intellectual and artistic pursuits.[51] In other words, the patient was to be distracted from ruminating on illness or work and redirecting self-absorbed thoughts to other external interests. Mitchell was convinced that much of the symptoms experienced by these difficult cases was brought about by excessive rumination and attention to those symptoms. Essentially, patient behaviors were reinforced by the attention and sympathy given by family and friends.

In order to prevent muscle wasting or atrophy associated with long periods of bed rest, a regimen of daily muscular therapy was initiated along with the rest. Both massage and electrical stimulation of voluntary muscles were employed. This part of the therapy posed the greatest challenge to Mitchell's thinking when he first conceived of the rest cure. He firmly believed that absolute rest was necessary for recovery and yet exercise of lax muscles was also needed.[52] The massage and electricity represented a passive exercise of the muscles. Mitchell's long-time assistant and associate indicated that Mitchell was well aware of the "harmfulness of too long or too monotonous rest measures."[53] Mitchell (1898) wrote of these measures, "From a restless life of irregular hours, and probably endless drugging, from hurtful sympathy and over-zealous care, the patient passes to an atmosphere of quiet, to order and control, to the system and care of a thorough nurse, to an absence of drugs, and to simple diet. The result is always at first, whatever it may be afterwards, a sense of relief, a remarkable and often a quite abrupt disappearance of many of the nervous symptoms."[54] The rest of the treatment involved a gradual resumption of physical activity and normal diet. "A major, and vital element in these convalescent procedures is the retraining of long disused muscles. Motor intelligence is an important part of general intelligence," wrote Mitchell's assistant.[55]

Finally, the detoxification element to the cure was undoubtedly responsible for much of the improvement wrought by the therapy. The daily amount and diversity of medications, almost all psychoactive, must have wreaked havoc on the nervous system. A variety of stimulants and depressants had usually been prescribed throughout the course of the illness prior to the initiation of the rest cure and included opium (usually in the form of laudanum), morphine, alcohol, lithium or potassium bromide, chloral hydrate, cannabis, arsenic, cocaine, and so on. Paregorics and emetics to purge the gastrointestinal tract were also employed; hydrotherapy and electric shocks were sometimes used to stimulate the nervous system. All in all, prior to undergoing the rest cure, these patients were likely overmedicated and probably dependent upon many of these drugs. Mitchell noted that typically the patient was able to be weaned off the addictive drugs during the first week of treatment.[56]

The Yellow Wall-Paper

This brings us to *The Yellow Wall-Paper*, a quasi-autobiographical short story published in 1892 by social activist Charlotte Perkins Gilman. For decades following its publication, it was considered a classic in the supernatural fiction genre. It was rediscovered in 1973 by scholars during the emerging Feminist movement, who seized upon the work as a now classic example of how a male-dominated medical system crushes the spirit of the heroine, ultimately driving her insane. It continues to be an important work in the canon of feminist literature.[57]

The story is told by an unnamed female narrator who has been brought to a rented house in the country by her physician husband. She reveals that the purpose was to provide rest and fresh air for "temporary nervous depression."[58] Their bedroom is the former nursery upstairs overlooking the garden. The narrator is immediately repulsed by wall's "repellent, almost revolting" yellow color.[59] Over time she becomes obsessed with the wallpaper and, though forbidden to engage in any activity such as writing, secretly journals about her reactions to the wallpaper. First, she perceives the figure of eyes in its pattern but eventually she sees a woman "creeping about" within the wallpaper.[60] Then she is convinced that the woman is able to escape periodically from the wallpaper and actually creeps around outside. Eventually, the narrator begins to strip off the wallpaper to help other women escape the wallpaper and by the end of the story, she herself has become one of those trapped women. This brief synopsis does not do justice to the intricacies of the plot nor the rich depiction of a gradual descent into insanity.

So what, then, is the connection with Silas Weir Mitchell? First, in the story, he is named directly. The narrator is threatened with being taken to see Dr. Mitchell if she does not get better during this enforced retreat.

More importantly, 20 years after writing the story, Gilman states her own reasons when she published "Why I Wrote *The Yellow Wall-Paper*" in 1913.[61] In the article, she reveals that she had been a patient of Mitchell's in 1887 due to a "nervous breakdown tending to Melancholia."[62] The Rest Cure had been applied for a month and when discharged from his care, Mitchell supposedly advised her to "live as domestic a life as far as possible," to "have but two hours' intellectual life a day," and "never to touch pen, brush or pencil again as long as I lived."[63] She claimed to have followed his instructions faithfully for three months until it drove her to the point of "utter mental ruin" and subsequently, the purpose of writing the story was to "save people from being driven crazy."[64] In her autobiography, published 40 years after the original story, she restated that the purpose of the story was to target Weir Mitchell explicitly in order to make him reconsider the merits of the rest cure.[65] Furthermore, she goes on to say in the article that a friend of hers had sent the story to Mitchell and upon reading it, he altered his Rest Cure from that point forward.[66] She not only believed that the Rest Cure had worsened her symptoms but that the cause of neurasthenia itself was the societal constraints placed upon women, in effect forcing them to lead nonproductive lives.

However, there are a couple of troubling things about her self-professed assertion and the modern-day assumption that the treatment described in the short story is an accurate representation of Mitchell's Rest Cure. A careful analysis of the treatment described in the story in fact bears little resemblance to the Rest Cure, as prescribed by Mitchell in his own words. The only similarity is the removal from her home to a new environment. However, a critical departure from treatment protocol—and a circumstance explicitly contraindicated by Mitchell—is that the narrator and husband share the bedroom. Besides an intimate family member being present, there is no evidence of the neutral, "kind but firm" nurse stranger that was recommended by Mitchell. Furthermore, it is clear from the text that the narrator has no enforced bed rest nor daily muscles treatments and milk diet. There are simply no other particular points of similarity for the specific therapy that are apparent. The similarities, instead, arise more from the culturally-embedded assumptions the underlie the doctor-patient dynamic that are firmly entrenched in perceptions of appropriate gender roles.[67]

It is also likely that Gilman suffered from what now would be considered post-partum depression and not neurasthenia.[68] Her deep melancholia followed the birth of her daughter. There are suggestions of that condition in *The Yellow Wall-Paper* with comments of a recent birth as well. Gilman's bouts of neurasthenia appeared to be triggered when confronted with returning to traditional family duties. From her own account, the Rest Cure therapy was initially apparently successful: declaring her cured, Mitchell discharged her from his care. It was Mitchell's post-therapy advice and conservative views towards women's roles that she ultimately found objectionable and drove her to despair.[69]

Finally, according to a thorough examination by Dock in her critical 1988 edition of the story published for its centennial, there is no evidence to Gilman's claim that Mitchell altered his therapy because of her story.[70] We only have her word for it based on memories 20 and 40 years, respectively, after her illness. She also alters statements and wordings in her drafts that suggest the issue is not so clear-cut and that she was perhaps engaging in revisionist history. There is no evidence among Mitchell's own papers and correspondence to suggest a change in his methods or attitudes regarding the Rest Cure, nor is there any evidence that his reputation at the time up to his death was negatively affected by the publication. It is only with the modern reading and scholarship that Mitchell has been vilified as the embodiment of male oppression.

It now seems that neurasthenia was strongly gendered as a uniquely American disorder, not in its manifestation but in the medical beliefs regarding cause and treatment. Males suffered as much, if not more, from neurasthenia than females.[71] For men, however, the prototype neurasthenic patient was an exhausted office worker and the prescribed treatment was exercise and travel! Only in cases where the male patient was a physical laborer did Mitchell recommend the Rest Cure to restore the nervous system.

Mitchell's Rest Cure was founded less deliberately upon patriarchal views intending to keep the status quo for women as it was simply representing an approach recognizing the psychology of the patient role, the current state of understanding of the nervous system, the effects of overmedication, and the dangers of anorexia. In fact, he did not distinguish between men and women on certain features of neurasthenia, especially in regard to the overmedication and the excessive rumination of the neurasthenic patient: "for to think too much about their disorders is, on the whole, one of the worst things which can happen to man or woman, and wholesome self-attention is difficult, nay, impossible to command without help from a personally-uninterested mind outside of oneself."[72] It is interesting to note that some aspects of his cure are now considered mainstream in the treatment of certain disorders, such as anorexia. Mitchell anticipated elements of the new psychotherapies with his focus on the mind-body connection and the interaction of patient and physician. Some of the Rest Cure's negative connotation is based on some scholars' reconstruction of events to fit a feminist narrative in the absence of evidence and applying the fallacy of presentism to neurology instead of examining the treatments in historical context.

Perhaps now we can more readily see the contributions of the Rest Cure and its foreshadowing to modern psychology without the emotional entanglement of the quasi-fictional account in *The Yellow Wall-Paper.* Mitchell's descriptions of patient behavior match contemporary concepts in Health Psychology, such as the role of secondary gains and compliance behaviors. Secondary gains refer to the relative and usually unacknowledged advantages given by the sick role.[73] The escape from responsibility, decision-making, and obligations can provide negative reinforcement for the patient. The characteristic rumination in depression has been studied by many contemporary researchers and has influenced cognitive therapies.[74] Mitchell's understanding of the interaction of stress and physical health and illness was also an important contribution—decades prior to the groundbreaking work of Hans Selye. Long before the General Adaptation Syndrome was outlined by Selye in the 1970s, Mitchell was describing the now familiar stages of alarm, resistance, and exhaustion.[75] These stages represent the bodily reaction to chronic stress: first, the sympathetic nervous system reacts with the classic "fight-or-flight" response; next, the body tries to adapt to the chronic stress because the acute sympathetic nervous system response cannot be sustained as such; finally, with no relief from the stress, the organism collapses.

Finally, I am not suggesting that Gilman's story should be discounted because it does not faithfully represent the Rest Cure. Whether or not the details of the therapy as described were accurate depictions of Mitchell's therapy or even Gilman's own personal experience, the narrative reflects the author's phenomenological perception of that experience. The story has value alone in teaching us about Victorian gender roles and the socially-constructed nature of a male-dominated society. However, the Rest Cure should also be evaluated on its own merits: in the context of an over-medicated society and near epidemic proportions of neurasthenia, Mitchell utilized the current understanding of neurological function at the time to bring together a systematic treatment of both mind and body.

Conclusion

THE PRECEDING CHAPTERS span a variety of neuroscience topics that weave together certain thematic elements in the field. The topics from a contemporary neuroscience standpoint may appear on the surface to be eclectic in nature, roughly categorized as dealing with overall brain function, psychoactive drugs and altered states of consciousness, and then applications of neuroscience. In examining the historical development of those same topics, however, the interconnections become more readily apparent.

One goal in writing this book was to explicate the links between current and past notions of nervous system action. Quite unintentionally, several chapters proved to have overlapping theories and concepts, major players, and points of commonality. With reflection, this happenstance is not as surprising as one might initially think, however. Some of the key scientists who made significant contributions in the history of neuroscience were eclectic in the scope of their own studies and scientific contributions, not limiting themselves to the very narrow academic specializations that exist today. Likewise, many were scientist-philosophers who posed general theories that permeated the field as a whole, influencing future research and development in a variety of areas, and those ideas are still recognizable in some form today.

In some instances, the topics themselves cut across more than one area of investigation. For example, in several chapters, neurasthenia appears as a relevant issue either as a causal factor or an effect of nervous disturbance. This was the name of a common nervous system pathology and was used as a diagnostic category for many decades. The term itself reflected the predominant theory of nervous system physiology at the time and so was discarded when new theories prevailed. However, the basic symptomology both preceded the terminology and continues today under the umbrella category of anxiety disorders. The interrelationship of nervous system and mental life—normal and abnormal—is both undisputed yet often avoided in neuroscience.

Phrenology came up again and again within several chapters. Here a controversial pseudoscience gave rise to many applications so that many of its tenets were widely disseminated to the public and in popular discourse. The principles that were actually accepted by the scientific community (e.g., the brain being the seat of the mind) have been lost in the overall discrediting of the practice. Thus, when one traces the lineage of certain topics, one is bound to find some connection with the phrenological science of the 18th and 19th centuries. As one of the earlier attempts to map the brain according to function, there will inevitably be remnants of phrenology when looking at the brain's role with music, passion, sleep, and character. The underlying assumption of structural localization of psychological abilities is still debated today and the visual methodologies of imaging share some similarity at times with the appeal of the phrenological contours of the mind. The imaging is now based on a more sound scientific basis but the underlying assumption that mental activity can be manifested visually can be directly traced to that earlier movement.

Another theme to emerge in this book considers the role of expert knowledge and disciplinary boundaries. Sometimes the boundaries are fixed and even manipulated to preserve professional territory, as in the case of the Viagra story. The problem of sexual impotence was firmly wrested from the psychological domain (e.g., psychiatry, psychology, sexology) by the urologists and pharmaceutical camps. A similar example is seen in the study of sleep, particularly sleep disorders. In other instances, the interdisciplinary nature of modern neuroscience has led significant breakthroughs in our understanding of physiology (e.g., the role of the gut in the nervous system functioning). Professional boundaries continue to be guided by underlying assumptions of dualism (mind-body separation) or monism (mind-body unity), although one can argue that modern neuroscience is perforce more holistic in nature—at least at the organismal level of analysis. History shows us that treating problems like drug and alcohol addiction and mental disorders using only a pharmaceutical approach, for example, are likely to fail overall.

A noteworthy observation concerning the history of various topics within this book is that the assumptions regarding the relationship of mind and body even for a single phenomenon have shifted back and forth at various times. This has been demonstrated by the various theories of sleep and the actions of psychoactive drugs. With the disciplinary compartmentalization that occurred with the rise of specialized areas in medicine, such as neurology, the once holistic medical approach to psychological and neurological disorders gave way to a separation of powers, so to speak. One of the consequences of the Freudian revolution was to move many pathologies, formerly considered neurological, firmly into the psychiatric realm. While perhaps unintended by Freud, a consequence was more entrenched separation of mind and body even permeating the bureaucracy of the modern health insurance industry which for decades has differentiated coverage of physiological versus psychological pathologies. Dualism is still pervasive in popular discourse.

In some instances, the historical links to contemporary theories are discontinuous, punctuated by long periods of inactivity, and so the thread between them is less visible. This is certainly the case with the current opiate epidemic. To some extent, scientific amnesia is the culprit: a tendency to focus only on the most current research findings at the expense of earlier work.[1] While often there is no need for an historical perspective, overlooking the past can potentially lead to situations with research considered ground-breaking yielding results that are suspiciously akin to findings from centuries ago. More importantly, a thorough examination of past successes and failures can create new insights in approaching particular problems. Even when the results are the same, methodologies considered as acceptable science in each era have helped to build a more complete picture of the phenomenon under investigation.

However, using the latest technology does not in and of itself insure greater validity of the overall findings. Specifically, today verifying a psychological phenomenon through brain imaging cerebral activity is somehow considered a more valid methodology compared with electrical recording and clinical case studies, not to mention patient self-report yet it does not take a brain scan to tell a doctor that a patient is drug-addicted or depressed. Depending on the research question being asked, older methods might, in fact, be superior or yield richer data. Modern neural imaging has indeed been an important advance in the technology of understanding brain activity but like all technologies, has its limitations and constraints. In the future, it is conceivable that particular neural signatures will be indicative of pain or addiction in ways that quantify variable subjective experiences.[2] But caution should be observed when those quantitative measures are given more credence than the patient's reported experience. Yes, some people will exaggerate pain for a variety of reasons that have little to do with their injury or condition but there still remain substantial individual differences in pain perception that means that patients at both extremes of the pain dimension will not be well treated on the basis of a statistical average applied to a brain scan.

Another theme in this book is the propensity of brain science in its various forms to be victimized by quackery. While quackery has by no means been limited to applications regarding the nervous system, there is a plethora of examples regarding fringe medicine and the brain. This is evident in several chapters and is not confined just to historical accounts. Contemporary neuroscience is sometimes distorted to provide a foundation for modern quackery as well, such as with detoxification programs. Sometimes the quackery is a deliberate fraud, as often the case with purported fringe medicinal cure-alls, such as Keeley's bi-chloride of gold. Sometimes, perhaps more interestingly, the quackery is an emergent phenomenon originally based in sincere beliefs stemming from the scientific knowledge of the day that lingers on after new discoveries have overtaken the scientific discourse. Phrenology was once considered legitimate science, for example, but was discarded once it was disproven with empirical evidence. The practice of applied phrenology, however, remained in the public eye for decades, long after being discounted by the scientific community.

Finally, can the juxtaposition of historical and contemporary accounts help elucidate what the nervous system actually does? In simple terms, the nervous system is the interface between the body and the external environment; it takes in sensory information from the external world and provides the means for responding to the environment. And yet, there are machines—most definitely without nervous systems—that can do the same thing! This observation has historically given rise to the concept of dualism: mind and body separation. The body, along with its nervous system, is considered more akin to a machine. This simple definition does not capture the essence of the nervous system's role in adaptation, however. The potential to change with experience, to avoid aversive stimuli, and to guide appetitive behaviors that promote survival, are all key aspects of nervous system action. Machines with artificial intelligence using sophisticated algorithms are capable of these actions as well, although one can certainly argue the latter teleological premise of promoting survival.

The nervous system, especially as evidenced by the brain, is also characterized by its ability to integrate information beyond simple input-output loops such as the case of a simple reflex. Thus far, we have still not strayed far from the realm of machinery, however. It is no accident that brain metaphors have included the "switchboard" or "information processor." But perhaps the most important aspect of the brain is the creation of an internal reality distinct from the external environment. Being able to distinguish oneself from the environment as an individual sentient being is perhaps one of the defining acts of the human nervous system. This is addressed in some theories of personality development, where the Self must be differentiated from the Other (cf. theories of Allport, Baldwin, Klein, Rogers).[3] This differentiation occurs during early infant development, a time during which the brain is undergoing massive neural pruning and maturation. A newborn baby is conscious yet presumably lacks a coherent sense of selfhood—an accomplishment that surely is accompanied by some necessary and sufficient functioning of the brain.

Contemporary neuroscience has done well in quests for localization: discerning where in the brain certain functions originate. Yet even then, the research literature is replete with contradictory findings and multiple answers to the same question. It is apparent that there are relatively few functions that can be isolated to a singular structure in the brain but at least for some phenomenon a more or less consistent picture has emerged. Again, the available technology limits the ability to "see" brain activity holistically. The plasticity of the brain enables both individual differences in response to environmental triggers and recovery in the event of assaults to the nervous system.

The "holy grail" in neuroscience continues to be the search for the neural basis of consciousness. In the words of neuroscientist Susan Greenfield, "How the 'water' of objective brain events is transformed into the 'wine' of subjective consciousness" is the seminal question.[4] Is consciousness singular and unitary or is it a multitudinous process? If it resides in the brain, and there certainly has been consensus since the late 18th and early 19th centuries that it does, is it found in one particular location or many? Is it a thing or a process? What gives rise to the Self, the awareness of existence—clearly more than simply taking sensory information in or producing a motor response. In many cases, simple reflexes need only an intact spinal cord to house the synapse of a sensory and motor neuron—no brain necessary. Whereas we can study altered states of consciousness, such as sleep or being under the influence of hallucinogenic drugs, the everyday, moment-to-moment experience of normal consciousness is still poorly understood. It is a problem that has vexed neuroscientists and philosophers alike.

Contemporary cognitive neuroscientists talk about the "Binding Problem": How does a unified conscious experience emerge from the different sensory experiences being processed in parallel in the brain?

Bottom-up processing leads to sensory integration from which holistic percepts are formed. While we know something of how that bottom-up processing might work—for example, the simple and complex neurons that respond to individual features of visual images, such as straight lines at various fixed orientations—much less is known about how those elements of visual experience are unified to become a comprehensible image.

The study of consciousness was one the critical topics of study by the first psychologists. Until the mid–18th century, studying the mind scientifically was thought by philosophers and scientists alike to be an impossible task. However, a new methodology derived from physics with the study of sensation revealed a path forward to investigate a mental phenomenon (i.e., perception). This led to the development of the scientific field of experimental psychology in contrast to the mental philosophy that had been previously called psychology. The first psychologists in the last decades of the 19th century studied mental activities, such as perception, volition, memory, attention and consciousness itself. Countless experiments were performed to examine the contents of consciousness and its function for the human condition.

With the ascent and then dominance of behaviorism as the theoretical framework for psychological experimentation emerging during the early decades of the 20th century, those topics related to consciousness waned in popularity. The chief objection to the study of consciousness arose from the methodologies utilizing introspection.[5] Such subjective analogs accompanying behavior were considered epiphenomena and superfluous to the experiment. Later, the famous metaphor of the mind as a "Black Box" emerged.[6] One could measure variables corresponding to stimuli that affected an organism (variables coming into the box) and one could measure outcome or response variables (variables emitted from the box) but what happened in between the two events or inside the box was unobservable.

It is fair to say that this state of affairs dominated experimental psychology until the so-called "cognitive revolution" in the 1960s and onward. Behaviorism had yielded significant contributions to our understanding of human and animal behavior but complaints about the "empty organism," humans devoid of the very characteristics that made them uniquely human, gave rise to new ways of knowing with both the cognitive perspective (coming from the experimental psychology tradition) and humanistic perspective (coming from the clinical/counseling psychology traditions). The shift was gradual at first but represented an unmistakable return by experimental psychology to the very concepts associated with its beginning. Improvements in experimental design methodologies and statistical analyses allowed for the acceptance of introspection, now combined with other concomitant measures of attention and consciousness. At the same time, developments in physiological psychology, cellular biology, and other related subfields, carried on with discoveries of neural communication and biochemical underpinnings of many basic processes of the nervous system and elements of behavior. Donald Hebb's theory of reverberating synaptic circuits as the physiological mechanism for learning, proposed in 1949 in the absence of empirical evidence at the intracellular level, has been affirmed by the newly discovered mechanisms of long-term potentiation.[7]

One cautionary note from my work is about the tendency to research in disciplinary silos. Because neuroscience is, by definition, an interdisciplinary field, there is some inherent protection against the isolationism of the past. Yet how and from where will unifying theories emerge? The contemporaneous developments in neuroscience have resulted in a focused program of study, cognitive neuroscience. Cognitive neuroscience has combined the research on mental processes, such as attention, memory, and consciousness, with the study of the nervous system to understand the underlying mechanisms of those phenomena. Indeed, of all the myriad academic disciplines that contribute to the field of neuroscience, cognitive neuroscience has the potential for being the unifying discipline because of its focus at the organismal level. The Nobel Prize–winning psychologist Roger Sperry stated, "The basic change is to a different paradigm for interlevel causal determinism. Instead of the traditional assumption that everything is determined exclusively from below upward, we posit also a reciprocal downward determinism. The combined 'double-way' or 'doubly determinate' model gives science a whole new way of perceiving, explaining, and understanding ourselves and the entire national order in a true Kuhnian worldview paradigm shift."[8]

To some extent, the juxtaposition of neuroscience past and present reveals "everything old is new again." Discussions on neuroscience topics, such as sleep, music and the brain, and personality types, have been recorded for centuries but the multi-disciplinary nature of neuroscience has sometimes meant that those past discussions occurred in isolation from today's conversations. An historical perspective can illuminate connections and the underlying assumptions of contemporary research problems. Conversely, employing an ahistorical approach not only does not potentially give prior credit where due, but can miss the important past discussions that led to either new discoveries or dead ends and so carry on to new ways of thinking. For example, learning about the prior struggles with opiate addiction can enrich the discussion of the present epidemic and perhaps point the way towards possible solutions.

I expressed the concern early on with this project that I would not be able to keep up with new developments on so many fronts in the field of neuroscience. My fear was that some information here would be outdated or even rendered obsolete as a point of contemporary comparison. When I began writing the chapter on phrenology and fMRIs, for example, the imaging protocols were much more varied and, in some cases, questionable in terms of validity. While caution is still required in interpreting imaging studies, there is now a greater appreciation in the field of the necessity of standardization in settings, statistical manipulations, and analysis. Newer technologies are also beginning to take hold. Likewise, as I wrote the chapter on opiates, the public was largely unaware of the brewing nascent opiate crisis. As I finish this book, that crisis is in full swing and someone would have to be very isolated indeed to remain unaware of the issue. My sincere hope is that by the time you are reading this, that crisis will have abated and fatal overdoses will be on a downward path.

A word of caution to the reader is in order—the history of neuroscience shows us that few of the research questions asked today are particularly new. The terminology and the level of analysis may change but the fundamental assumptions surrounding our understanding of brain and behavior have been guided by ways of knowing established centuries ago. I am not suggesting that those questions have been answered satisfactorily in the past but instead, that a bit of humility and acknowledgment of lessons learned from the past can assist rather than hinder the path towards new discoveries.

Chapter Notes

Preface

1. Abi-Rached & Rose, 2010.

Introduction

1. Insel, 2006.
2. Jacyna & Casper, 2012.
3. Littlefield & Johnson, 2012.
4. Finger, 2001.
5. Cooter, 2014.
6. E.g., Churchland, 2013; Johnston & Parens, 2014; Legrenzi & Umiltá, 2011; Satel & Lilienfeld, 2012; Taylor, K., 2012.
7. Abi-Rached & Rose, 2010; Rose & Abi-Rached, 2013.
8. Aserinsky & Kleitman, 1953.
9. Loe, 2004.
10. Skinner, 1964.
11. Gilman, 2000/1892.

Chapter 1

1. Clarke, E., & Jacyna, 1987; Young, 1968.
2. Satel & Lilienfeld, 2012.
3. Satal & Lilienfeld, 2012; Uttal, 2001; Wager, Hernandez, Jonides, & Lindquist, 2007; February 2016 issue of *Scientific American Mind*.
4. Clarke, E., & Jacyna, 1987.
5. Gall, 1822–1826/1935; Spurzheim, 1815.
6. Feinberg & Farah, 2000.
7. Harrington, 1987, p. 7.
8. Colbert, 1997.
9. Bell, C., 1811/1966, p. 2.
10. Finger, 2000.
11. Finger, 2000; Spurzheim, 1833.
12. Finger, 2000.
13. Davies, 1955, p. 8.
14. Gieryn, 1999, p. 125.
15. Cited in Harrington, 1987, p. 8.
16. Finger, 2000.
17. Tomlinson, 2005.
18. Tomlinson, 2005.
19. E.g., Cooter, 2001, Davies, 1955; Gieryn, 1999.
20. Combe, 1828.
21. Adapted from Cooter, 2001; Sewall, 1839.
22. Finger, 2001; Harrison, 1825; Sewall, 1839.
23. Glover, 1851.
24. Clarke, E., & Jacyna, 1987; Finger, 2001.
25. Feinberg & Farah, 2000; Harrington, 1987.
26. Fancher, 1979.
27. Finger, 2000; 2001.
28. Finger, 2000; 2001.
29. James, 1890, p. 28.
30. Glover, 1851.
31. Young, 1968, p. 255.
32. Fancher, 1979, p. 103.
33. Bakan, 1966.
34. Colbert, 1997, p. 23.
35. Van Wyhe, 2004a.
36. Walsh, 1974, p. 448.
37. Stern, M., 1971.
38. Stern, M., 1971.
39. C.f. "Phrenology," 1824; "Review," 1826.
40. Colbert, 1997.

41. Walsh, 1974.
42. Van Wyhe, 2004b.
43. "Report," 1842, p. 703.
44. Ridgway, 1993.
45. Cooter, 2001.
46. C.f. numerous articles by Elliotson in the 1837 *Lancet*.
47. Glover, 1851.
48. "Phrenology," 1824.
49. Broussais, 1836, p. 417.
50. "Natural Language," 1843, p. 423.
51. Zeidler & Sadler, 2000.
52. Wager & Lindquist, 2011.
53. Wager, Hernandez, & Lindquist, 2009.
54. Wager et al., 2009.
55. Wager et al., 2009.
56. Wager et al., 2009.
57. Bennett, Baird, Miller, & Wolford, 2009.
58. Bennett, Wolford and Miller, 2009.
59. Wager et al., 2009.
60. Wager et al., 2009.
61. Wager et al., 2009, p. 158.
62. Wager et al., 2009.
63. Wager et al., 2009.
64. Satel and Lilienfeld, 2013.
65. Bakan, 1966.
66. Drew, Vo, & Wolfe, 2013.
67. Lyons, 2009.
68. Reiser, 2009, p. 25.
69. Reiser, 2009, p. 26.
70. Reiser, 1993.
71. Reiser, 1993, p. 832.
72. Reiser, 1993.
73. Combe, 1828, pp. 57–65.
74. Reiser, 1993.
75. Reiser, 1993, p. 843.
76. Reiser, 1993, p. 842.
77. Tomlinson, 2005.
78. Joyce, 2008.
79. Golinski, 1998.
80. Golinski, 1998, p. 153.
81. Joyce, 2008, p. 61.
82. Joyce, 2008, p. 46.
83. Benschop & Draaisma, 2000.
84. Benschop & Draaisma, 2000, p. 6.
85. Benschop & Draaisma, 2000, p. 2.
86. Golinski, 1998.
87. Weisberg, Keil, Goodstein, Rawson, & Gray, 2008.
88. Fernandez-Duque, Evans, Christian, & Hodges, 2015.

Chapter 2

1. Gazzaniga, Bogen, & Sperry, 1962.
2. Zatorre, Evans & Meyer, 1994.
3. Puente, 2012.
4. Harris, 1999.
5. Corballis, 2007.
6. Edwards, 1979.
7. Gazzaniga, 2015.
8. Gazzaniga, 2000.
9. Gazzaniga, 2000, p. 1294.
10. Gazzaniga, 2000.
11. Eliassen, Baynes, & Gazzaniga, 2000.
12. Luck, Hillyard, Mangun, & Gazzaniga (1989).
13. Gazzaniga, 2005.
14. Geschwind & Galaburda, 1987.
15. Corballis, 2007.
16. Funnell, Corballis, & Gazzaniga, 1999.
17. Corballis, Inati, Funnell, Grafton, & Gazzaniga, 2001.
18. Benton, 1975.
19. Watson, N., & Breedlove, 2016.
20. Benton, 1975, p. 35.
21. Lokhorst, 1982.
22. Lokhorst, 1982, pp. 34–35.
23. Harrington, 1987.
24. Clarke, E., & Jacyna, 1987.
25. Harrington, 1987.
26. Harrington, 1987.
27. Harrington, 1987, p. 15.
28. Bichat, 1800/2015.
29. Harrington, 1987.
30. Watson, 1836.
31. Clarke, B., 1987.
32. Holland, 1852.
33. Wigan, 1844a.
34. Wigan, 1844b, p. 39.
35. Wigan, 1844b.
36. Clarke, B., 1987.
37. Wigan, 1844a; 1844b.
38. Wigan, 1844b, p. 40.
39. Clarke, B., 1987.
40. Wigan, 1844b.
41. Wigan, 1844b, p. 41.
42. Wigan, 1844a; 1844b; Winslow, 1849.
43. Winslow, 1849, p. 509.
44. Winslow, 1849, p. 506.
45. Wigan, 1844b.
46. Wigan, 1844a; 1844b.
47. Wigan, 1844b, p. 40.
48. Clarke, B., 1987, p. 32.

49. Holland, 1852.
50. Holland, 1852, p. 187.
51. Clarke, B., 1987.
52. Finger, 2001.
53. Harrington, 1987.
54. Harris, 1999.
55. Harrington, 1987.
56. Harrington, 1987; Jacyna, 1982.
57. Harrington; 1987; Harris, 1999.
58. Finger, 2000; Harrington, 1987.
59. Finger, 2000.
60. Finger, 2000.
61. Harrington, 1987.
62. Harrington, 1987.
63. Cited in Harrington, 1987, p. 58.
64. Harrington, 1987.
65. Harrington, 1987, p. 73.
66. Harris, 1999.
67. Harrington, 1987, p. 78.
68. Harrington, 1987; Harris, 1999.
69. Harris, 1999, p. 21.
70. Benton, 1972; 1975; Ireland, 1891.
71. Jackson, 1868, p. 208.
72. Harris, 1999, p. 20.
73. Jackson, 1868, p. 208, 358.
74. Benton, 1972.
75. Harris, 1999, p. 24.
76. Harrington, 1987, p. 85.
77. Harrington, 1987.
78. Harrington, 1987.
79. Harrington, 1987.
80. Harrington, 1987.
81. Harris, 1984.
82. Harris, 1984; Ireland, 1886.
83. Ireland, 1880.
84. Lombroso, 1903.
85. Lombroso, 1903, p. 442.
86. Lombroso, 1903, p. 444.
87. Watson, N., & Breedlove, 2016.
88. Corballis, 2007, p. 16.
89. Harrington, 1987, p. 125.
90. Harrington, 1987.
91. Harrington, 1987, p. 109.
92. Bruce, 1895.
93. Bruce, 1895, p. 54.
94. Harrington, 1987, p. 120.
95. Brown-Séquard, 1890, p. 627.
96. Brown-Séquard, 1890, p. 628.
97. Brown-Séquard, 1890, p. 632.
98. Maudsley, 1889.
99. Maudsley, 1889, p. 162.
100. Harrington, 1987.
101. Harrington, 1987, p. 126.
102. Maudsley, 1889.
103. Maudsley, 1889, p. 179.
104. Maudsley, 1889, p. 163.
105. Harrington, 1987.
106. Clarke, B., 1987; Harrington, 1987.
107. Harrington, 1987.
108. Harrington, 1987.
109. Harrington, 1987, p. 151.
110. Harrington, 1987.
111. Harrington, 1987, p. 257.
112. Lashley, 1950.
113. Corballis, 2007.
114. Clarke, B., 1987; Harrington, 1987.
115. Corballis, 2007.
116. Gazzaniga, 2000; 2005.
117. Harrington, 1987, p. 228.
118. Harris, 1999.
119. E.g., Bassett & Gazzaniga, 2011; Koch & Marcus, 2014; Tononi & Koch, 2008.
120. Pinto et al., 2017, p. 6.
121. Puente, 1993.

Chapter 3

1. Jouvet, 1967; 1999.
2. Watson, N., & Breedlove, 2016.
3. Dement & Kleitman, 1957; Watson, N., & Breedlove, 2016.
4. Jouvet, 1965; 1999.
5. Aserinsky & Kleitman, 1967.
6. Watson, N., & Breedlove, 2016.
7. Chokroverty & Billiard, 2015; Stevens, 2015.
8. Stevens, 2015, p. 4.
9. Stevens, 2015.
10. Hobson, 2005.
11. Hobson, 2005, p. 1255.
12. Watson, N., & Breedlove, 2016.
13. Stevens, 2015.
14. Saper, Scammell, & Lu, 2005, p. 1259.
15. Stevens, 2015.
16. Watson, N., & Breedlove, 2016.
17. Saper et al., 2005.
18. Chokroverty & Billiard, 2015.
19. Arnold & Vogel, 2007.
20. Williams, 2005.
21. Dannenfeldt, 1986; Ekirch, 2015.
22. Rather, 1968, p. 337.
23. Rather, 1968.
24. Yanjiao et al., 2015.
25. Loza, 2015.
26. Kumar, 2015.
27. Thorpy, 2011; 2015.

28. Cited in Arnold & Vogel, 2007, p. 23.
29. Descartes, 1972.
30. Damjanovic, Milovanovic, & Trajanevic, 2015, p. 399.
31. Damjanovic et al., 2015.
32. Cogan, 1612; Cheyne 1724.
33. Kroker, 2007, p. 63.
34. Willis, 1692.
35. Thorpy, 2015; Williams, 2005.
36. Hartley, D., 1749.
37. Thorpy, 2015.
38. Cabanis & Mora, 1981.
39. Kroker, 2007.
40. MacNish, 1842.
41. MacNish, 1842, p. 1.
42. MacNish, 1842, p. 26.
43. Hammond, 1869.
44. Kroker, 2007.
45. Hammond, 1869.
46. Hammond, 1869, p. 42.
47. Hammond, 1869, p. 63.
48. Hammond, 1869, p. 98.
49. Kroker, 2007.
50. Williams, 2005.
51. Hammond, 1869.
52. Kroker, 2007.
53. Kroker, 2007.
54. Freud, 1900.
55. Kroker, 2007, p. 129.
56. Williams, 2005.
57. Kroker, 2007, p. 187.
58. Chokroverty & Billiard, 2015; Ishimori, 1909; Piéron, 1913.
59. Kroker, 2007, p. 157, 162.
60. Kroker, 2007, p. 169.
61. Kroker, 2007.
62. Crichton-Miller, 1930.
63. Aserinsky & Kleitman, 1953.
64. Dement, 1993.
65. Kroker, 2007.
66. Dement, 1993.
67. Aserinsky & Kleitman, 1953.
68. Dement, 1958; Dement & Kleitman, 1957.
69. Kroker, 2007.
70. Gottesman, 2009; Jouvet, 1965.
71. Kroker, 2007, p. 324.
72. Jouvet, 1967; Kroker, 2007.
73. Williams, 2005.
74. Williams, 2005, p. 150.
75. Loza, 2015.
76. Kumar, 2015.
77. Ancoli-Israel, 2015.
78. Ekirch, 2015.
79. Cogan, 1612, p. 235.
80. Willis, 1692.
81. C.f. Clarke, C., 1897; Crichton-Miller, 1930; Hammond, 1869; MacNish, 1842; Rudolf, 1922; Schulz & Salzarulo, 2015; Symonds, 1925.
82. Gordon, S., 2015, p. 426.
83. Umanath, Sarezky & Finger, 2011.
84. Snelders, Kaplan & Pieters, 2006.
85. Schulz & Salzarulo, 2015.
86. Snelders et al., 2006.
87. Zisapel, 2012.
88. Schwartz & Woloshin, 2009.

Chapter 4

1. Graziano & Johnson, 2015.
2. Rauscher, Shaw & Ky, 1993.
3. Albusac-Jorge & Giménez-Rodríguez, 2015.
4. O'Kelly, Fachner and Tervaniemi, 2016, p. 1.
5. Helmholtz, 1863/1954.
6. Lenoir, 1994.
7. Stumpf, 2012/1911.
8. De la Motte-Haber, 2012.
9. Koelsch, 2013.
10. Koelsch, 2013.
11. Koelsch, 2013.
12. Koelsch, 2013.
13. Koelsch, 2013, p. 93.
14. Koelsch, 2013.
15. Zatorre & McGill, 2005.
16. Koelsch, 2013.
17. Zatorre & McGill, 2005.
18. Zatorre & McGill, 2005, p. 314.
19. Karmonik et al., 2016.
20. Margulis, 2008.
21. Margulis, 2008, p. 160.
22. Rauscher et al., 1993.
23. Margulis, 2008.
24. C.f. Schellenberg, 2013.
25. Schellenberg, 2013.
26. Blood & Zatorre, 2001.
27. Kennaway, 2010.
28. Rousseau, 2014, p. 31.
29. Kennaway, 2010.
30. Kennaway, 2010, p. 401.
31. Kennaway, 2010, p. 402.
32. Cheyne, 1733.
33. Whytt, 1768.
34. Browne, 1729; Kennaway, 2010; 2012.
35. Kennaway, 2012, p. 24.

36. Browne, 1729.
37. Kennaway, 2012.
38. Kennaway, 2012.
39. Brown, J., 1795.
40. Kennaway, 2012, p. 34.
41. Kennaway, 2010; 2012.
42. Kennaway, 2012.
43. Kennaway, 2012, p. 51.
44. Kennaway, 2012.
45. Rousseau, 2014, p. 32.
46. Rousseau, 2014.
47. C.f. Hanslick, 1891.
48. Bashwiner, 2011.
49. Bashwiner, 2011, p. 246.
50. Hadlock, 2000, p. 509.
51. Hadlock, 2000, p. 524.
52. Hadlock, 2000, p. 524.
53. Kennaway, 2012, p. 45.
54. Hadlock, 2000; Kennaway, 2012.
55. Hadlock, 2000.
56. Sykes, 2014.
57. Kennaway, 2012.
58. Kennaway, 2012, p. 64.
59. Kennaway, 2012.
60. Kennaway, 2012.
61. Johnson, 2011; Kenneway, 2012.
62. Johnson, 2011.
63. Johnson, 2011, p. 20.
64. Johnson, 2011, p. 24.
65. Kennaway, 2012.
66. Critchley, 1977; Scott, 1977.
67. Scott, 1977, p. 356.
68. Stern, J., 2015; Swaminathan, 2008.

Chapter 5

1. Kalant, 1987; Karch, 2006.
2. Aschenbrandt, 1883.
3. Bentley, 1880.
4. Karch, 2006.
5. Kalant, 1987.
6. Freud, 1968.
7. Karch, 2006.
8. Rushton & Steinberg, 1963.
9. Karch, 2006.
10. Freud, 1884/1984.
11. Karch, 2006, p. 109.
12. Gay, 1988.
13. Freud, 1968, p. 76.
14. Freud, 1885; 1884/1984; 1887/1963.
15. Thornton, 1986.
16. Karch, 2006.
17. E.g., Breger, 2000; Gay, 1988; Karch, 2006; Thornton, 1986.
18. Karch, 2006.
19. Freud, 1885.
20. Thornton, 1986.
21. Erlenmeyer, 1885, p. 427.
22. Erlenmeyer, 1887.
23. E.g., Mattison, 1887.
24. Freud, 1887/1963.
25. Masson, 1985; Thornton, 1986.
26. McGuire, 1974.
27. Davenport-Hines, 2001; Karch, 2006.
28. Freud, 1968.
29. Freud, 1900.
30. James, 1890.
31. Barzun, 1983, p. 23.
32. Perry, 1948, p. 362.
33. Croce, 1999, p. 308.
34. James, 1902.
35. Brown, E., 1993.
36. Brown, E., 1993, p. 566.
37. Brown, E., 1993, p. 574.
38. Brown, E., 1993, p. 575.
39. Taylor, E., 1996; Bjork, 1983.
40. Simon, 1998, p. 141.
41. Simon, 1998, p. 141.
42. Skrupskelis & Berkeley, 1992a.
43. Skrupskelis & Berkeley, 1992b.
44. Mitchell, 1896.
45. Skrupskelis & Berkeley, 1992b, p. 403.
46. Earnest, 1950.
47. James, 1882; 1898; 1902; 1910.
48. James, 1882, p. 208.
49. Taylor, E., 1996, p. 92.
50. Rylance, 2000, p. 7.
51. Ellis, 1897; Mitchell, 1896.
52. Grosskurth, 1980.
53. Ellis, 1897; 1898.
54. Ellis, 1897.
55. Ellis, 1898.
56. Ellis, 1898.
57. Ellis, 1898, p. 140.
58. Ellis, 1898.
59. Ellis, 1898, p. 140.
60. Ellis, 1898, p. 141.
61. Ellis, 1911, p. 23.
62. Ellis, 1940.
63. Ellis, 1898; 1911; Sachs, 2012.
64. Ellis, 1911, p. 28.
65. Sacks, 2012, p. 124.
66. Santouse, Howard, & ffytche, 2000.
67. Booth, 2003, pp. 155–156.
68. Grosskurth, 1980.
69. Ellis, 1898.
70. Ellis, 1897.

71. Ellis, 1902, p. 1542.
72. Ellis, 1898, p. 141.
73. Zieger, 2008.
74. Ellis, 1898, p. 141.
75. Jacyna, 1981.
76. Jacyna, 1981, p. 110.
77. Smith, 1993, p. 79.
78. Daston, 1978.
79. Smith, 1993.
80. Mitchell, 1896, p. 1625.

Chapter 6

1. Grob, Bossis & Griffiths, 2013, p. 294.
2. Kroll & Bachrach, 2005.
3. Cohen, 1970.
4. Lewin, 1998/1927.
5. Bromberg & Tranter, 1943.
6. E.g., Bromberg & Tranter, 1943.
7. Schultes, 1938.
8. Mooney, 1896; Mitchell, 1896; Prentiss & Morgan, 1896; Ellis, 1897.
9. Hoffer & Osmond, 1967.
10. Knauer & Maloney, 1913.
11. Hermle, Spitzer & Gouzoulis, 1994.
12. Cited in Klüver, 1966/1928, p. viii.
13. Klüver, 1966/1928; Beringer, 1927.
14. Hofmann, 1970, p. 6.
15. C.f. Bandura, 1982.
16. Hofmann, 1970.
17. Hofmann, 1970, p. 3.
18. Hofmann, 1970.
19. Hofmann, 1970, p. 5.
20. Stevens, 1987.
21. Dyck, 2005.
22. Cohen, 1970, p. 34.
23. Craig & Kauffman, 2006.
24. Witt, 1954.
25. Osmond & Smythies, 1952.
26. Cohen, 1970.
27. Guttmann & Maclay, 1936.
28. Dyck, 2005, p. 56.
29. C.f. Carhart-Harris et al., 2014.
30. Passie, 1997.
31. Passie, 1997.
32. E.g., Kast, 1967.
33. Abramson, 1967.
34. Cohen, 1970.
35. Dyck, 2005.
36. Dyck, 2005, p. 108.
37. E.g., Hoffer, 1970; Smart & Storm, 1964.
38. Hoffer, 1970.
39. Cohen, 1970.
40. Cohen, 1970.
41. Dyck, 2005.
42. Dyck, 2005.
43. Cohen, 1970.
44. C.f. Stevens, 1987.
45. Stevens, 1987.
46. Stevens, 1987.
47. Stevens, 1987.
48. Cohen, 1960.
49. Stevens, 1987.
50. Stevens, 1987.
51. Cohen, 1970; Stevens, 1987.
52. Oram, 2014, p. 222.
53. Novak, 1977.
54. Lee & Shlain, 1985.
55. Lieberman, 1962, p. 11.
56. Lieberman, 1962, p. 11.
57. Lieberman, 1962.
58. Lieberman, 1962.
59. Lee & Shlain, 1985.
60. Lee & Shlain, 1985.
61. Lieberman, 1962.
62. Lee & Shlain, 1985; Lieberman, 1962.
63. Lee & Shlain, 1985.
64. Lee & Shlain, 1985.
65. Lee & Shlain, 1985.
66. Oram, 2014.
67. Oram, 2014, pp. 231–232.
68. Oram, 2014.
69. Oram, 2014.
70. Oram, 2014.
71. Grinspoon & Bakalar, 1975; Oram, 2014.
72. Dyck, 2005.
73. Pletscher & Ladewig, 1994.
74. E.g., Hermle, Spitzer, & Gouzoulis, 1994; Richter, 1994.
75. Brown, D., 2007/2008.
76. Kurzweil, 1995.
77. Tupper, Wood, Yensen, & Johnson, 2015.
78. E.g., Brown, D., 2007/2008; Grob et al., 2013; Moreno, Wiegand, Taitano, & Delgado, 2006.
79. C.f. Green, 2008.
80. Gaddum & Vogt, 1956; Green, 2008; Halberstadt, 2015.
81. Vollenweider et al., 1997.
82. Carhart-Harris et al., 2012; Carhart-Harris et al., 2016.

Chapter 7

1. Acker, 2003, p. 53.
2. CDC, 2012.
3. Bernstein, 2015.
4. NIDA, 2016.
5. NIDA, 2016.
6. O'Donnell, Halpin, Mattson, Goldberger, & Gladden, 2017.
7. Kolata & Cohen, 2016.
8. Kochanek, Murphy, Xu, & Arias, 2017.
9. C.f. CDC, 2017.
10. Nelson, 2015b.
11. Nelson, 2015a.
12. Portenoy & Foley, 1986; Porter, J., & Jick, 1980.
13. Quinones, 2015.
14. Quinones, 2015, p. 95.
15. Foreman, 2014; Quinones, 2015.
16. Foreman, 2014.
17. Foreman, 2014, p. 127.
18. Grill, 2013.
19. CDC, 2012.
20. Baldisseri, 2007; Merlo, Singhakant, Cummings & Cottler, 2013; SAMSA, 2014.
21. E.g., de Grandpre, 2006; Greene, 2007; Healy, 2012.
22. Musto, 1999, p. 69.
23. Watson, N., & Breedlove, 2012.
24. Mischel, 2015.
25. Eysenck, 1957.
26. Berridge, K., 2007; Robinson, T., & Berridge, K., 1993.
27. Aurin, 2000, p. 414.
28. Berridge, V., 1999.
29. Berridge, V., 1977; Musto, 1999; Peters, 1981.
30. Christison, 1832.
31. Christison, 1832, p. 132.
32. Foxcroft, 2007.
33. Milligan, 2005; Levinstein, 1878.
34. Berridge, V., 1978; Hodgson, 1999.
35. Clarke, J., 1808.
36. Berridge, V., 1999; Parssinen, 1983.
37. Parssinen, 1983, p. 42.
38. Parssinen, 1983, p. 43.
39. Berridge, V., 1978; Hodgson, 1999.
40. Parssinen & Kerner, 1980.
41. Parssinen & Kerner, 1980, p. 291.
42. Acker, 2003.
43. Crothers, 1902.
44. Levinstein, 1878.
45. Parssinen, 1983.
46. Parssinen, 1983.
47. Foxcroft, 2007.
48. Levinstein, 1878.
49. Parssinen, 1983.
50. Auren, 2000.
51. Foxcroft, 2007, pp. 113–114.
52. E.g., Kerr, 1894.
53. Parssinen, 1983.
54. Foxcroft, 2007, p. 125.
55. Acker, 2002.
56. Kandall, 1996.
57. Acker, 2002; Oppenheim, 1991.
58. Foxcroft, 2007.
59. Acker, 2002.
60. Berridge, V., 1999.
61. Beard, 1880; Foxcroft, 2007.
62. Berridge, V., 1978.
63. Acker, 2002; Auren, 2000.
64. Berridge, V., 1978, p. 460.
65. Auren, 2000, p. 419.
66. Auren, 2000.
67. Courtwright, 1982.
68. E.g., De Grandpre, 2006.
69. Aurin, 2000; Musto, 1999.
70. Straus, Ghitza, & Tai, 2013.

Chapter 8

1. Gregory, 2007.
2. Gregory, 2007, p. 70.
3. Haubrich, 1993.
4. Gregory, 2007.
5. Glover, 1851, p. 37.
6. Trotter, 1807.
7. Cheyne, 1733.
8. Gregory, 2007; Trotter, 1807.
9. Gregory, 2007.
10. Gregory, 2007, p. 75.
11. Shryock, 1931.
12. Young, 1992.
13. Young, 1992, p. 173.
14. Young, 1992.
15. Helfand, 1974.
16. Helfand, 1974.
17. Helfand, 2002.
18. Helfand, 2002.
19. Helfand, 1974, p. 102.
20. Helfand, 2002.
21. Helfand, 1974.
22. Helfand, 1974, p. 106.
23. Porter, 2003b, p. 317.
24. Porter, 2003b, p. 317.
25. Reprinted in Moat, 1831.

26. Wakley, 1836b, p. 57.
27. Porter, 2000.
28. Porter, 2000, p. 203.
29. Helfand, 1974.
30. Helfand, 1974; 2002; Porter, 2003b.
31. Helfand, 1974.
32. Helfand, 1974.
33. "Morison's Hygeian Pills," 1836, p. 207.
34. "Morison's Hygeian Pills," 1836.
35. Wakley, 1836a.
36. Labatt & Stokes, 1833, p. 238.
37. Moat, 1832, p. 85.
38. Wakley, 1836c.
39. Wakley, 1836c, p. 92.
40. Wakley, 1836d.
41. Helfand, 1974.
42. Porter, 1989.
43. Jameson, 1961, p. 196.
44. Severance, Prandovszky, Castiglione, & Yolken, 2015.
45. E.g., Bucknill, 1857; Woodward, 1850; Workman, 1863.
46. Workman, 1863, p. 46.
47. Beard, 1884.
48. Beard, 1884, p. 250.
49. Mitchell, 1898.
50. Mitchell, 1898.
51. Mitchell, 1898.
52. Haas, 1924.
53. Abel, 2010.
54. Levenstein, 1988, p. 86.
55. Haas & Haas, 1950.
56. Abel, 2010, p. 100.
57. Abel, 2010.
58. Levenstein, 1988, p. 204.
59. Rayner, 2017.
60. www.gapsdiet.com.
61. Rea, Dinan & Cryan, 2016.
62. Rea et al., 2016.
63. Rea et al., 2016.
64. Mayer, 2016; Yarandi, Peterson, Treisman, Moran, & Pasricha, 2016.
65. Mayer, 2016.
66. E.g., Severance, et al., 2015; Severance, Yolken, & Eaton, 2016; Yarandi et al., 2016.
67. Severance et al., 2016.
68. Severance et al., 2016.
69. Callaghan, 2017.

Chapter 9

1. E.g., Loe, 2004; Marshall, B., 2002.
2. Loe, 2004, p. 10.
3. Brunton, Gilman & Goodman, 2008; Fishman, 2007.
4. Loe, 2004.
5. Fishman, 2007.
6. Loe, 2004.
7. Fishman, 2007.
8. Loe, 2004, p. 10.
9. Hare, 1962.
10. Mumford, 1992, p. 35.
11. Stolberg, 2000.
12. Stolberg, 2000.
13. Tissot, 1832.
14. Mumford, 1992.
15. Hare, 1962.
16. Stolberg, 2000.
17. Stolberg, 2000.
18. Engelhardt, 1974.
19. Mumford, 1992.
20. Stolberg, 2000, p. 40.
21. Hall, L., 1992, p. 365.
22. Hall, L., 1992, p. 366.
23. Hall, L., 1992, pp. 369–370.
24. Engelhardt, 1974, p. 235.
25. Beard, 1900.
26. Tissot, 1832, p. 51.
27. Hare, 1962.
28. Hall, L., 1992.
29. C.f. Engelhardt, 1974.
30. Conrad & Schneider, 1980, p. 180.
31. Beard, 1900.
32. Beard, 1900, p. 93.
33. Beard, 1900, p. 94.
34. Beard, 1900, p. 123.
35. Beard, 1900, p. 125.
36. Beard, 1900, p. 125.
37. Breuer & Freud, 1895/1955.
38. Mumford, 1992.
39. Conrad & Schneider, 1980.
40. Mumford, 1992, p. 44.
41. Mumford, 1992, p. 47.
42. Conrad & Schneider, 1980.
43. Beard, 1900; Mumford, 1992.
44. Beard, 1900; Mumford, 1992.
45. Acton, 1875; Hall, L., 1992.
46. Peterson, 1986.
47. Peterson, 1986.
48. Hall, L., 1992, p. 365.
49. Hall, L., 1992, p. 369.
50. Hall, L., 1992.
51. Hopkins, 1883.
52. Stall, 1897.
53. Hall, L., 1992, p. 372.
54. Hall, L., 1992, p. 373.

55. Hare, 1962; Robinson, W., 1912.
56. Hall, L., 1992.
57. Hall, G., 1904.
58. E.g., Anders, 1900; Osler, 1901.
59. Schuster, 2006.
60. Hirschbein, 2000.
61. Brown-Séquard, 1889.
62. Marshall, B., & Katz, 2002; Matfin, 2010.
63. Hirshbein, 2000.
64. Matfin, 2010.
65. Hirshbein, 2000.
66. Hirshbein, 2000.
67. Marshall, B., & Katz, 2002, p. 53.
68. E.g., Casper, 1906; Fuller, 1900.
69. Casper, 1906, p. 553.
70. E.g., Ostojic, 2006.
71. E.g., Ballenger, 1914; Greene & Brooks, 1912; Morton, 1919.
72. E.g., Keyes, 1923; Young & Davis, 1926.
73. E.g., Hinman, 1935.
74. C.f. Eisendrath & Rolnick, 1938; Herman, 1938.
75. Lowsley & Kirwin, 1940, p. 293.
76. Lowsley & Kirwin, 1940, p. 293.
77. C.f. McCrea, L., 1948.
78. C.f. Marshall, V., 1956.
79. Parton, 1960.
80. Parton, 1960. p. 234.
81. Loe, 2004.
82. Loe, 2004, p. 30.
83. Loe, 2004, p. 38.
84. Loe, 2004, p. 39.
85. Schulte, 2015.
86. Hartley, H., & Tiefer, 2003.
87. Jaspers et al., 2016.
88. Loe, 2004.

Chapter 10

1. DeYoung et al., 2010.
2. Smith, 1997.
3. Smith, 1997.
4. Smith, 1997.
5. Rousseau, 2007.
6. Rousseau, 2007.
7. Kagan & Snidman, 2004; Rousseau, 2007.
8. Kagan, 2013.
9. Parssinen, 1974.
10. Parssinen, 1974, p. 3.
11. Parssinen, 1974; Young, 1968.
12. Bell, C., 1811/1966, p. 3.
13. Hartley, L., 2001.
14. Parssinen, 1974, p. 7.
15. Clarke, E., & Jacyna, 1987.
16. Spurzheim, 1815.
17. Clarke, E., & Jacyna, 1987; Smith, 1997.
18. Smith, 1997, p. 213.
19. Hartley, L., 2001, p. 1.
20. Lavater, 1789.
21. Hartley, L., 2001, p. 2.
22. Feinberg & Farrah, 2000.
23. Hartley, L., 2001, p. 12.
24. Porter, 2003a.
25. Hartley, L., 2001, p. 40.
26. Hartley, L., 2001, p. 40.
27. Porter, 2003a.
28. Porter, 2003a.
29. Lavater, 1789.
30. Hartley, L., 2001.
31. Hartley, L., 2001.
32. Hartley, L., 2001, p. 16.
33. Hall, M., 1837.
34. Clarke & Jacyna, 1987; Logan, 1997.
35. Logan, 1997.
36. Kempf, 1918.
37. Berman, 1921.
38. Smith, 1992.
39. Sherrington, 1904.
40. Smith, 1992; 1997.
41. Maher & Maher, 1994.
42. Maher & Maher, 1994, p. 75.
43. Smith, 1992.
44. Smith, 1992.
45. Eysenck, 1947; 1967.
46. Smith, 1997.
47. Maher & Maher, 1994.
48. Jung, 1971.
49. Smith, 1992, p. 205.
50. Eysenck, 1997; 2000.
51. E.g., Robinson, D., 2001.
52. Maher & Maher, 1994.
53. Tanner, 1947.
54. E.g., Enke, 1933.
55. Vertinsky, 2007, p. 299.
56. Farber, 1938.
57. Allport, 1937.
58. McCrae, R., & Costa, 1987.
59. John & Srivastava, 1999.
60. John & Srivastava, 1999, p. 129.
61. E.g., De Moor et al., 2012; Dumont, 2010; Jang, Livesley & Vernon, 1996.
62. De Moor et al., 2012.
63. Dumont, 2010.
64. E.g., De Young, 2013; Hermes, Hage-

mann, Naumann, & Walter, 2011; Rothbart, Derryberry & Posner, 1994.

65. DeYoung, 2013.
66. Watson, N., & Breedlove, 2016.
67. Smillie & Wacker, 2014.
68. DeYoung et al., 2010.
69. DeYoung et al., 2010, p. 822.
70. DeYoung, 2013.
71. DeYoung, 2013, p. 8.
72. Hermes et al., 2011.
73. Rothbart et al., 1994.
74. Abram & DeYoung, 2017.
75. DeYoung et al., 2010, p. 822.
76. Ebstein, 2006.
77. Kagan, 1984; Kagan & Snidman, 2004.
78. Kagan, 2013.

Chapter 11

1. Bynum, 1984.
2. Jay, 2010.
3. Shortt, 1986; Appel, 2008.
4. Cheyne, 1733; Shapin, 2003.
5. Jones, 1700.
6. Stewart, 1905.
7. Berridge, V., 1999.
8. Rotunda, 2007, p. 51.
9. Collins, 1903; Gordon, M., 1904.
10. Berridge, V., 1999.
11. Levinstein, 1878.
12. Musto, 1999.
13. Musto, 1999.
14. Helfand, 1996.
15. Levinstein, 1878.
16. Berridge, V., 1999.
17. C.f., "Drug Treatments," 1909.
18. Helfand, 1996.
19. Helfand, 1996.
20. "The Composition," 1909, p. 909.
21. "The Composition," 1909.
22. Cramp, 1921.
23. E.g., "The Composition," 1909; Cramp, 1921.
24. Helfand, 1996; White, 1998.
25. Helfand, 1996.
26. White, 1998.
27. Helfand, 2002.
28. White, 1998.
29. White, 1998, p. 51.
30. Keeley, 1892b.
31. Dormandy, 2012; Kandall, 1996.
32. Helfand, 2002, p. 185.
33. White, 1998.
34. Tracy, 2005, p. 86.
35. White, 1998.
36. Dormandy, 2012.
37. Bannister, 1891–92.
38. White, 1998.
39. White, 1998.
40. White, 1998.
41. Helfand, 2002; Kandall, 1996.
42. White, 1998.
43. Dormandy, 2012.
44. White, 1998.
45. White, 1998.
46. Cited in White, 1998, p. 52.
47. White, 1998, p. 52.
48. Keeley, 1892b.
49. Keeley, 1892b, pp. 34–35.
50. Helfand, 2002.
51. Helfand, 2002.
52. White, 1998.
53. White, 1998.
54. Cited in Freed, 2007, p. 114.
55. Squire, 1892.
56. White, 1998.
57. White, 1998.
58. White, 1998, p. 54.
59. White, 1998.
60. White, 1998, p. 55.
61. White, 1998, p. 57.
62. White, 1998.
63. "Report," 1892.
64. "Report," 1892, p. 107.
65. "Report," 1892, p. 107.
66. C.f. Edmunds, 1892; Tracy, 2005; Usher, 1892.
67. Keeley, 1892a.
68. White, 1998.
69. Usher, 1892.
70. White, 1998.
71. White, 1998.
72. White, 1998.
73. White, 1998.
74. Tracy, 2005; White, 2008.
75. Tracy, 2005.
76. Helfand, 1996.
77. Keeley, 1892b.
78. Maisto, Galizio, & Connors, 2015.
79. McCabe, 2000.
80. Centers for Disease Control, 2013.
81. Helfand, 1996.
82. Berridge, K., & Robinson, T., 2016.
83. Humphreys, 2012.
84. E.g., Raby, 2000.
85. Ling et al., 2011.
86. Humphreys, 2011.

87. Humphreys, 2011.
88. Humphreys, 2011.
89. Ling et al., 2011.
90. Nestor, 2016.
91. Nestor, 2016.

Chapter 12

1. Gilman, 1892/2000.
2. Denizet-Lewis, 2017; Geraldo, 2017; O'Regan, 2017.
3. Bandelow & Michaelas, 2015.
4. Bandelow & Michaelas.
5. Shimada-Sugimoto, Otawa, & Hettema, 2015.
6. Bandelow & Michaelas, 2015.
7. Cited in Tone, 2009.
8. Cited in Tone, 2009.
9. Herzberg, 2009.
10. Herzberg, 2009.
11. Herzberg, 2009.
12. Beard, 1869; Beard, 1881.
13. Tone, 2009.
14. Rosenberg, 1997, p. 107.
15. Gosling, 1987; Rosenberg, 1997.
16. C.f. Sicherman, 1977.
17. Corn, 2004, p. 1.
18. Rosenberg, 1997.
19. Freedman, 1987, p. 2.
20. Beard, 1880, p. 115.
21. Taylor, J., 1929.
22. Rosenberg, 1997.
23. Rosenberg, 1997.
24. Schuster, 2006.
25. Mitchell, 1898.
26. Cocke, 1894, Ch. 13.
27. De Paula Ramos, 2003.
28. Freud, 1895/1955.
29. Schuster, 2006; Tone, 2009.
30. Mitchell, 1896.
31. Skrupskelis & Berkeley, 1992, p. 403.
32. Burr, 1929.
33. Blount, Openshaw, & Todd, 1940; Burr, 1929; Walter, 1970.
34. Mitchell, 1872.
35. Bell, W., 1987.
36. Sicherman, 1977.
37. Mitchell, 1872, p. 11.
38. Mitchell, 1872; Walter, 1970.
39. Mitchell, 1872.
40. Van Wyck Good, 1993, p. 51.
41. Mitchell, 1872.
42. Mitchell, 1872, p. 9.
43. Mitchell, 1872, p. 39.
44. Mitchell, 1872, p. 40.
45. Mitchell, 1872, p. 49.
46. Mitchell, 1872, p. 42.
47. Taylor, J., 1929, p. 591.
48. Baynes, 1888.
49. Lawlor, 2012; Mitchell, 1885.
50. Mitchell, 1898.
51. Mitchell, 1898.
52. Mitchell, 1904.
53. Taylor, J., 1929, p. 591.
54. Mitchell, 1898, p. 61.
55. Taylor, J., 1929, p. 592.
56. Mitchell, 1898, p. 124.
57. Dock, 1998.
58. Stetson, 1898/1998, p. 29.
59. Stetson, 1898/1998, p. 31.
60. Stetson, 1898/1998, p. 35.
61. Gilman, 1913.
62. Gilman, 1913, p. 271.
63. Gilman, 1913, p. 271.
64. Gilman, 1913, p. 271.
65. Gilman, 1963.
66. Dock, 1998.
67. Mitchell, 1888; Oppenheim, 1991; Stiles, 2013.
68. Schuster, 2006.
69. Gilman, 1963.
70. Dock, 1998.
71. Schuster, 2006.
72. Mitchell, 1888, p. 134.
73. Fishbain, 1994.
74. C.f. Nolen-Hoeksema & Davis, 1999.
75. Selye, 1978.

Conclusion

1. Leary, 1985.
2. Wager et al., 2013.
3. Allport, 1955; Baldwin, 1895; Klein, 1959/1984; Rogers, 1959.
4. Greenfield, 2016, p. 28.
5. Watson, J., 1913.
6. C.f. Skinner, 1964.
7. Bliss & Gardner-Medwin, 1973; Hebb, 1949.
8. Sperry, 1995, p. 506.

Bibliography

Abel, E. K. (2010). The rise and fall of celiac disease in the United States. *The Journal of the History of Medicine and Allied Sciences, 65,* 81–105.

Abi-Rached, J.M., & Rose, N. (2010). The birth of the neuromolecular gaze. *History of the Human Sciences, 23,* 11–36.

Abram, S. V., & DeYoung, C. G. (2017). Using personality neuroscience to study personality disorder. *Personality Disorders: Theory, Research, and Treatment, 8,* 2–13.

Abramson, H. (Ed.). (1967). *The use of LSD in psychotherapy and alcoholism.* New York: Bobbs-Merrill.

Acker, C. J. (2002). *Creating the American junkie.* Baltimore: Johns Hopkins University Press.

Acker, C. J. (2003). Take as directed: The dilemmas of regulating addictive analgesics and other psychoactive drugs. In M. L. Meldrum (Ed.). *Opioids and pain relief: A historical perspective* (pp. 35–55). Seattle: IASP Press.

Acton, W. (1857). *The function and disorders of the reproductive organs.* Philadelphia: Lindsay & Blakiston.

Albusac-Jorge, M., & Giménez-Rodríguez, F. J. (2015). Citation index and scientific production on the neuroscience of music: A bibliometric study. *Psychomusicology, 25,* 416–422.

Allport, G. W. (1937). *Personality: A psychological interpretation.* New York: Holt, Rinehart & Winston.

Allport, G. W. (1955). *Becoming: Basic consideration for a psychology of personality.* New Haven: Yale University Press.

Ancoli-Israel, S. (2015). Sleep in the Biblical period. In S. Chokroverty & M. Billiard (Eds.), *Sleep Medicine* (pp. 35–42). New York: Springer.

Anders, J.M. (1900). *A text-book of the practice of medicine* (4th ed.). London: W.B. Saunders.

Appel, J. M. (2008). "Physicians are not bootleggers": The short, peculiar life of the medicinal alcohol movement. *Bulletin of the History of Medicine, 82,* 355–386.

Arnold, K., & Vogel, K. (2007). Introduction. In K. Arnold & K. Vogel, *Sleeping and dreaming* (pp. 19–25). London: Black Dog.

Aschenbrandt, T. (1883). Die physiologische Wirkung und die Bedeutung des Cocains. *Deutsche Medizinische Wochenschrift, 9,* 730.

Aserinsky, E., & Kleitman, N. (1953). Regularly occurring periods of eye motility, and concomitant phenomena during sleep. *Science, 118,* 273–274.

Aurin, M. (2000). Chasing the dragon: The cultural metamorphosis of opium in the United States, 1825–1935. *Medical Anthropology Quarterly, 14,* 414–441.

Bakan, D. (1966). The influence of phrenology on American psychology. *The Journal of the History of the Behavioral Sciences, 2,* 200–220.

Baldisseri, M.R. (2007). Impaired healthcare professionals. *Critical Care Medicine, 35*(2), S106–16.

Baldwin, J. M. (1895). *Mental development in the child and the race.* New York: Macmillan.

Ballenger, E. G. (1914). *Genito-urinary diseases and syphilis.* London: Butterworth.

Bandelow, B., & Michaelas, S. (2015). Epidemiology of anxiety disorders in the 21st century. *Dialogues in Clinical Neuroscience, 17,* 327–335.

Bandura, A. (1982). The psychology of chance encounters and life paths. *American Psychologist, 37,* 747–755.

Bannister, H. M. (1891–92). The bi-chloride of gold cure for inebriety. *American Journal of Insanity, 48,* 471–472.

Barzun, J. (1983). *A stroll with William James.* Chicago: University of Chicago Press.

Bashwiner, D. M. (2011). Lifting the foot: The neural underpinnings of the "pathological" response to music. In B. M. Stafford (Ed.), *A field guide to a new meta-field: Bridging the humanities-neurosciences divide* (pp. 239–266). Chicago: University of Chicago Press.

Bassett, D. S., & Gazzaniga, M. S. (2011). Understanding complexity in the human brain. *Trends in Cognitive Science, 15,* 200–209.

Baynes, D. (1888). *Auxiliary methods of cure.* London: Simpkin, Marshall.

Beard, G. (1869). Neurasthenia, or nervous exhaustion. *Boston Medical and Surgical Journal, 3,* 217–221.

Beard, G. (1880). *A practical treatise on nervous exhaustion (neurasthenia).* New York: William Wood.

Beard, G. (1881). *American nervousness: Its causes and consequences.* New York: G. P. Putnam & Sons.

Beard, G. M. (1884). *Sexual neurasthenia [Nervous exhaustion]* (1st ed.). New York: E. B. Treat.

Beard, G. M. (1900). *Sexual neurasthenia* (5th ed.). New York: E. B. Treat.

Bell, C. (1966). *Idea of a new anatomy of the brain.* [reprint]. London: Dawsons (Original work published 1811)

Bell, W. J. (1987). *The College of Physicians of Philadelphia.* Canton, MA: Science History.

Bennett, C. M., Baird, A. A., Miller, M. B., & Wolford, G. L. (2009). Neural correlates of interspecies perspective taking in the postmortem Atlantic salmon: An argument for multiple comparisons correction. Retrieved from http://prefrontal.org/files/posters/Bennett-Salmon-2009.pdf

Bennett, C. M., Wolford, G. L., & Miller, M. B. (2009). The principled control of false positives in neural imaging. *Social Cognitive and Affective Neuroscience, 4,* 417–422.

Benschop, R., & Draaisma, D. (2000). In pursuit of precision: The calibration of mind and machines in late nineteenth-century psychology. *Annals of Science, 57,* 1–25.

Bentley, W. H. (1880). *Erythroxylon Coca. Therapeutic Gazette, 1*(12), 350–351.

Benton, A. L. (1972). The "minor" hemisphere. *Journal of the History of Medicine, 27,* 5–14.

Benton, A. (1975). Historical development of the concept of hemispheric cerebral dominance. In S. F. Spicker, & H. T. Engelhardt (Eds.), *Philosophical dimentions of the neuro-medical sciences* (pp. 35–57). Boston: D. Reidel.

Beringer, K. (1927). *Das Meskalinrausch.* Berlin: Springer.

Berman, L. (1921). *The glands regulating personality.* New York: Macmillan.

Bernstein, L. (2015, December 18). Deaths from opiate overdoses set a record in 2015. *The Washington Post.* Retrieved from https://www.washingtonpost.com/news/to-your-health/wp/2015/12/11/deaths-from-heroin-overdoses-surged-in-2014/.

Berridge, K. C. (2007). The debate over dopamine's role in reward: The case for incentive salience. *Psychopharmacology, 191,* 391–431.

Berridge, K. C., & Robinson, T. E. (2016). Liking, wanting, and the incentive-sensitization theory of Addiction. *American Psychologist, 71,* 670–679.

Berridge, V. (1977). Opium and the historical perspective. *Lancet, 8028,* 78–80.

Berridge, V. (1978). Victorian opium-eating: Responses to opiate use in nineteenth century England. *Victorian Studies, 21,* 437–462.

Berridge, V. (1999). *Opium and the people: Opiate use and drug control policy in nineteenth and early twentieth cen-*

tury England. London: Free Association Press.

Bichat, X. F. E. (2015). *Recherches physiologique sur la vie et la mort.* London: Forgotten Books. (Original work published 1800.)

Bjork, D. W. (1983). *William James: The center of his vision.* Washington, D.C.: American Psychological Association.

Bliss, T. V. P., & Gardner-Medwin, A. R. (1973). Long-lasting potentiation of synaptic transmission in the dentate area of the unanaesthetized rabbit following stimulation of the perforant path. *Journal of Physiology, 232,* 357–374.

Blood, A. J., & Zatorre, R. J. (2001). Intensely pleasurable responses to music correlate with activity in brain regions implicated in reward and emotion. *Proceedings of the National Academy of Sciences, 98,* 11818–11823.

Blount, B. K., Openshaw, H. T., & Todd, A. R. (1940). The *Erythrophleum* alkaloids. Part I. Erythrophleine. *The Journal of the Chemical Society, 0,* 286–290.

Booth, M. (2003). *Cannabis: A history.* London: Bantam.

Breger, L. (2000). *Freud: Darkness in the midst of vision.* New York: John Wiley & Sons.

Breuer, J., & Freud, S. (1955). *Studies on Hysteria.* Trans. & Ed. J. Strachey in *The Standard Edition of the Complete Psychological Works of Sigmund Freud* (Vol. 2.). London: Hogarth Press. (Original work published 1985.)

Bromberg, W., & Tranter, C. L. (1943). Peyote intoxication: Some psychological aspects of the peyote rite. *Journal of Nervous and Mental Disease, 97,* 518–527.

Broussais, M. (1836, June 25). Lectures on phrenology. *The Lancet, 2* (669), 417–423.

Brown, D. J. (2007/2008). Psychedelic healing? *Scientific American Mind, 18*(6), 67–71.

Brown, E. M. (1993). Neurology and spiritualism in the 1870s. *Bulletin of the History of Medicine, 57,* 563–577.

Brown, J. (1795). *The elements of medicine.* London: W. & D. Treadwell.

Brown-Séquard, C. E. (1889). The effects produced on man by subcutaneous injections of a liquid obtained from the testicles of animals. *Lancet, 2,* 105–107.

Brown-Séquard, C. E. (1890). Have we two brains or one? *Forum, 9,* 627–643.

Browne, R. (1729). *Medicina musica; or, A mechanical essay on the effects of singing, musick, and dancing, on human bodies.* London: J. & J. Knapton.

Bruce, L. C. (1895). Notes of a case of dual brain action. *Brain, 18,* 54–65.

Brunton, L. L., Gilman, A., & Goodman, L. S. (2008). *Goodman & Gilman's manual of pharmacology and therapeutics.* New York: McGraw-Hill.

Bucknill, J. C. (1857). On the pathology of insanity. *The American Journal of Insanity, 14,* 29, 172, 254, 348.

Burr, A. R. (1929). *Weir Mitchell: His life and letters.* New York: Duffield.

Bynum, W. F. (1984). Alcoholism and degeneration in 19th century European medicine and psychiatry. *British Journal of Addiction, 79,* 59–70.

Cabanis, P. J. G., & Mora, G. (1981). *On the relations between the physical and moral aspects of man.* Baltimore: Johns Hopkins University Press.

Callaghan, B. L. (2017). Generational patterns of stress: Help from our microbes? *Current Directions in Psychological Science, 26,* 323–329.

Carhart-Harris, R. L., Erritzoe, D., Williams, T., Stone, J. M., Reed, L. J., Colasanti, A., ... & Nutt, D. J. (2012). Neural correlates of the psychedelic state as determined by fMRI studies with psilocybin. *Proceedings of the National Academy of Sciences, 109,* 2138–2143.

Carhart-Harris, R. L., Leech, R., Hellyer, P. J., Shanahan, M., Feilding, A., Tagliazucchi, E., ... & Nutt, D. (2014). The entropic brain: A theory of conscious states informed by neuroimaging research with psychedelic drugs. *Frontiers in Human Neuroscience, 8*(Article 20), 1–22. doi: 10.3389/fnhum.2014.00020.

Carhart-Harris, R. L., Muthukumaraswamy, S., Roseman, L., Kaelen, M., Droog, W., Murphy, K., ... & Nutt, D. J. (2016). Neural correlates of the LSD experience revealed by multimodal neuroimaging. *Proceedings of the National Academy of Sciences, 113,* 4853–4858.

Casper, L. (1906). *A text-book of genito-urinary diseases.* London: Rebman.

Centers for Disease Control (2012). CDC

grand rounds: Prescription drug overdoses, a U.S. epidemic. *Morbidity & Mortality Weekly Report, 61*(01), 10–13.

Centers for Disease Control (2013). Deaths and severe adverse events associated with anesthesia-assisted rapid opioid detoxification—New York City 2012. *Morbidity & Mortality Weekly Report, 62*(38), 777–780.

Centers for Disease Control (2017). Vital signs: Changes in opiate prescribing in the United States, 2006–2015. *Morbidity & Mortality Weekly Report,* 66(26), 697–704.

Cheyne, G. (1724). *An essay of health and long life.* London: G. Strahan.

Cheyne, G. (1733). *The English malady: Or a treatise of nervous diseases of all kinds, as spleen, vapours, lowness of spirits, hypochondriacal, and hysteric dystempers, etc.* London: G. Strahan.

Chokroverty, S., & Billiard, M. (2015). Introduction. In S. Chokroverty & M. Billiard (Eds.), *Sleep medicine* (pp. 1–9). New York: Springer.

Christison, R. (1832). Cases and observations in medical jurisprudence: Case X. On the effects of opium-eating on health and longevity. *Edinburgh Medical and Surgical Journal, 37,* 123–135.

Churchland, P.S. (2013). *Touching a nerve: Our brains, our selves.* New York: W. W. Norton.

Clarke, B. (1987). Arthur Wigan and the duality of the mind. *Psychological Medicine* (Supplement 17). (Ed. M. Shepherd). Cambridge: Cambridge University Press.

Clarke, C. K. (1897). A discussion on the treatment of insomnia. *British Medical Journal, II,* 853–866.

Clarke, E., & Jacyna, L. S. (1987). *Nineteenth-century origins of neuroscientific concepts.* Berkeley: University of California Press.

Clarke, J. (1808). Report from "the General Hospital near Nottingham." *Edinburgh Medical & Surgical Journal, 4,* 265–285.

Cocke, J. R. (1894). *Hypnotism: How it is done; Its uses and dangers.* New York: Arena.

Cogan, T. (1612). *The haven of health.* London: Malch Bradywood.

Cohen, S. (1960). Lysergic Acid Diethylamide: Side effects and complications. *Journal of Nervous and Mental Diseases, 130,* 30–39.

Cohen, S. (1970). *Drugs of hallucinations: The LSD story* (2nd ed.). St. Albans, UK: Paladin.

Colbert, C. (1997). *A measure of perfection: Phrenology and the fine arts in America.* Chapel Hill: University of North Carolina Press.

Collins, W. J. (1903). The institutional treatment of inebriety. *The British Medical Journal,* 2 (2236), 1204–1208. The composition of certain secret remedies. (1909). *The British Medical Journal, 1* (259), 909–911.

Conrad, P., & Schneider, J. W. (1980). *Deviance and medicalization: From badness to sickness.* St. Louis: C. V. Mosby.

Coombe, G. (1828). *The constitution of man.* Edinburgh: Neill.

Cooter, R. (2001). *Phrenology in America and Europe* (8 Volumes). London: Routledge/Thoemmes Press.

Cooter, R. (2014). Neural veils and the will to historical critique: Why historians of science need to take the neuro-turn seriously. *Isis, 105,* 145–154.

Corballis, M. (2007). The dual-brain myth. In S. Della Salla (Ed.), *Tall tales on the brain* (pp. 291–313. Oxford: Oxford University Press.

Corballis, M., Inati, S. J., Funnell, M. G., Grafton, S., & Gazzaniga, M. S. (2001). MRI assessment of spared fibers following callosotomy: A second look. *Neurology, 57,* 1345–1346.

Corn, W. M. (2004). Brain fag. In Z. Ross (Ed.). *Women on the verge: The culture of neurasthenia in nineteenth-century America* (pp. 1–5). Palo Alto: Stanford University Press.

Courtwright, D. T. (1982). *Dark paradise.* Cambridge: Harvard University Press.

Craig, G. W., & Kauffman, G. B. (2006). LSD: Catalyst for a pharmacological revolution. *The Chemical Educator, 11,* 421–426.

Cramp, A. J. (1921). Nostrums and quackery. *Journal of the American Medical Association, 76,* 9–17.

Crichton-Miller, H. (1930). *Insomnia: An outline for the practitioner.* London: Edward Arnold.

Critchley, M. (1977). Musicogenic epilepsy. In M. Critchley, & R. A. Henson (Eds.), *Music and the brain* (pp. 344–353). London: Wm. Heinemann Medical Books.

Croce, P. J. (1999). Physiology as the antechamber to metaphysics: The young William James's hope for a philosophical psychology. *History of Psychology, 2,* 302–323.

Crothers, T. D. (1902). *Morphinism and narcomanias from other drugs.* Philadelphia: W. B. Saunders.

Damjanovic, A., Milovanovic, S. D., & Trajanovic, N. N. (2015). Descartes and his peculiar sleep pattern. *Journal of the History of the Neurosciences, 24,* 396–407.

Dannenfeldt, K. H. (1986). Sleep: Theory and practice in the late Renaissance. *Journal of the History of Medicine and Allied Sciences, 41,* 415–444.

Daston, L. J. (1978). British responses to psycho-physiology, 1860–1900. *Isis, 69,* 192–208.

Davenport-Hines, R. (2001). *The pursuit of oblivion: A global history of narcotics 1500–2000.* London: Weidenfeld & Nicolson.

Davies, J. D. (1955). *Phrenology: Fad and science.* New Haven: Yale University Press.

De Grandpre, R. (2006). *The cult of pharmacology.* Durham: Duke University Press.

de la Motte-Haber, H. (2012). Carl Stumpf: Impulses towards a cognitive theory of music evolution. In D. Trippett (Ed.). *Origins of music* (pp. 3–15). Oxford: Oxford University Press.

Dement, W. C. (1958). The occurrence of low voltage fast electroencephalogram patterns during behavioral sleep in the cat. *Electroencephalography in Clinical Neurophysiology, 10,* 291–296.

Dement, W. C. (1993). The history of narcolepsy and other sleep disorders. *The Journal of the History of Neuroscience, 2,* 121–134.

Dement, W., & Kleitman, N. (1957). The relation of eye movements during sleep to dream activity: An objective method for the study of dreaming. *Journal of Experimental Psychology, 53,* 339–346.

De Moor, M. H. M., Costa, P. T., Terracianno, A., Krueger, R. F., de Geus, E. J. C., Toshiko, T., ..., Boomsma, D. I. (2012). Meta-analysis of genome-wide association studies for personality. *Molecular Psychiatry, 17,* 337–349.

Denizet-Lewis, B. (2017, October 11). Why are more American teenagers than ever suffering from severe anxiety? *The New York Times* [on-line edition]. Retrieved from https://www.nytimes.com/2017/10/11/magazine/why-are-more-american-teenagers-than-ever-suffering-from-severe-anxiety.html.

De Paula Ramos, S. (2003). Revisiting Anna O.: A case of chemical dependence. *History of Psychology, 6,* 239–250.

Descartes, R. (1972). *Treatise of man.* Trans. T. S. Hall. Cambridge: Harvard University Press. (Original work published 1664.)

DeYoung, C. G. (2013). The neuromodulator of exploration: A unifying theory of the role of dopamine in personality. *Frontiers in Human Neuroscience, 7* (Article 762). [On-line article], 26 pp. DOI: 10.3389/fnhum.2013.00762.

DeYoung, C. G., Hirsh, J. B., Shane, M. S., Papademetris, X., Rajievan, N., & Gray, J. R. (2010). Testing predictions from personality neuroscience: Brain structure and the Big Five. *Psychological Science, 21,* 820–828.

Dock, J. B. (Ed.). (1998). *Charlotte Perkins Gilman's "The yellow wall-paper" and the history of its publication and reception: A critical edition and documentary casebook.* University Park: Pennsylvania State University Press.

Dormandy, T. (2012). *Opium: Reality's dark dream.* New Haven: Yale University Press.

Drew, T., Vo., M. L. H., & Wolfe, J. M. (2013). The invisible gorilla strikes again: Sustained inattentional blindness in expert observers. *Psychological Science, 24,* 1848–1853.

Drug treatments for inebriety. (1909). *The British Medical Journal, 1,* 1072.

Dumont, F. (2010). *A history of personality psychology.* Cambridge: Cambridge University Press.

Dyck, E. (2005). Psychedic psychiatry: LSD and post–World War II medical experimentation in Canada. Unpublished doctoral dissertation, McMaster University.

Earnest, E. (1950). *S. Weir Mitchell: Novelist and physician.* Philadelphia: University of Pennsylvania Press.
Ebstein, R. P. (2006). The molecular genetic architecture of human personality: Beyond self-report questionnaires. *Molecular Psychiatry, 11,* 427–445.
Edmunds, J. (1892). "Cures" for inebriety. *Lancet, II,* 285.
Edwards, B. (1979). *Drawing on the right side of the brain.* New York: Tarcher.
Eisendrath, D. N., & Rolnick, H. C. (1938). *Urology* (4th ed.). Philadelphia: J. B. Lippincott.
Ekirch, A. R. (2015). Sleep medicine in the middle ages and the Renaissance. In S. Chokroverty & M. Billiard (Eds.), *Sleep Medicine* (pp. 63–67). New York: Springer.
Eliassen, J. C., Baynes, K., & Gazzaniga, M. S. (2000). Anterior and posterior callosal contributions to simultaneous bimanual movements of the hands and fingers. *Brain, 123,* 2501–2511.
Elliotson, J. (1837). On the ignorance of the discoveries of Gall, evinced by recent physiological writers. *The Lancet, 1,* 295–298.
Ellis, H. (1897, June). A note on the phenomena of mescal intoxication. *Lancet, I,* 1540–1542.
Ellis, H. (1898, January). Mescal: A new artificial paradise. *The Contemporary Review,* 140–142.
Ellis, H. (1902, May). Mescal: A study of a divine plant. *Popular Science Monthly,* 52–71.
Ellis, H. (1911). *The world of dreams.* London: Constable.
Ellis, H. (1940). *My life.* London: Wm. Heinemann.
Engelhardt, T. (1974). The disease of masturbation: Values and the concept of disease. *Bulletin of the History of Medicine, 48,* 234–248.
Enke, E. (1933). The affectivity of Kretschmer's constitution types as revealed in psycho-galvanic experiments. *Journal of Personality, 1,* 225–233.
Erlenmeyer, A. (1885). Erlenmeyer on the effect of cucaine [*sic*] in the treatment of morphinomania. *Journal of Mental Science, 31,* 427–428.
Erlenmeyer, A. (1887). *Die Morphiumsucht und ihre Behandlung* (3rd Ed.). Berlin: Heusser.
Eysenck, H. (1947). *Dimensions of personality.* New York: Praeger.
Eysenck, H. J. (1957). Drugs and personality. *The British Journal of Psychiatry, 103*(403), 119–131.
Eysenck, H. (1967). *The biological basis of personality.* Springfield, IL: Charles C. Thomas.
Eysenck, H. (1997). Addiction, personality and motivation. *Human Psychopharmacology, 12,* S79–S87.
Eysenck, H. (2000). Personality as a risk factor in cancer and coronary heart disease. In D. T. Kenny, J. G. Carlson, F. J. McGuigan, & J. L. Sheppard (Eds.), *Stress and health: Research and clinical applications* (pp. 291–318). Amsterdam: Harwood Academic Publishers.
Fancher, R. E. (1979). *Pioneers of psychology.* New York: W. W. Norton.
Farber, M. L. (1938). A critique and an investigation of Kretschmer's theory. *Journal of Abnormal and Social Psychology, 33,* 398–404.
Feinberg, T. E., & Farah, M. J. (2000). A historical perspective on cognitive neuroscience. In M. J. Farah, & T. E. Feinberg (Eds.). *Patient-based approaches to cognitive neuroscience* (pp. 3–20). Cambridge: MIT Press.
Fernandez-Duque, D., Evans, J., Christian, C., & Hodges, S. D. (2015). Superfluous neuroscience information makes explanation of psychological phenomena more appealing. *Journal of Cognitive Neuroscience, 27,* 926–944.
Finger, S. (2000). *Minds behind the brain.* New York: Oxford University Press.
Finger, S. (2001). *Origins of neuroscience* (reprint). New York: Oxford University Press.
Fishbain, D. (1994). Secondary gain concept: Definition problems and its abuse in medical practice. *American Pain Society Journal, 3,* 264–273.
Fishman, J. (2007). Viagra. In A. Tone, & E.S. Watkins (Eds.). *Medicating modern America* (pp. 229–252). New York: New York University Press.
Foreman, J. (2014). *A nation in pain.* New York: Oxford University Press.

Foxcroft, L. (2007). *The making of addiction.* Aldershot, UK: Ashgate.

Freed, C. R. (2007). Addiction medicine and addiction psychiatry in America: The impact of physicians in recovery on the medical treatment of addiction. *Contemporary Drug Problems, 34,* 111–136.

Freedman, A. M. (1987). Introduction. In F. G. Gosling (1987). *Before Freud: Neurasthenia and the American medical community, 1870–1910.* Urbana: University of Illinois Press.

Freud, S. (1885). Beitrag zur Kenntniss der Cocawirkung. *Wiener Medizinische Wochenschrift, 35,* 129.

Freud, S. (1984). Über coca. *Journal of Substance Abuse, 1,* 206–217. (Original work published 1884.)

Freud, S. (1900). *The interpretation of dreams.* London: Hogarth.

Freud, S. (1963). Craving for and fear of cocaine. In *The cocaine papers.* [Trans. L. A. Freisinger]. Vienna: Dunquin Press. (Original work published 1887.)

Freud, S. (1968). *Brautbriefe: Briefe an Martha Bernays aus den Jahren 1882–1886.* Frankfurt: Fisher Bücherei.

Fuller, E. (1900). *Diseases of the genito-urinary system.* London: Macmillan.

Funnell, M. G., Corballis, P. M., & Gazzaniga, M. S. (1999). A deficit in perceptual matching in the left hemisphere of a callosotomy patient. *Neuropsychologia, 37,* 1143–1154.

Gaddum, J. H., & Vogt, M. (1956). Some central actions of 5-hydroxytryptamine and various antagonists. *British Journal of Pharmacology, 11,* 175–179.

Gall, F. J. (1935). *On the functions of the brain.* [Trans. W. Lewis]. Boston: Marsh, Capen & Lyon. (Original work published 1822–1826.)

Gay, P. (1988). *Freud: A life for our time.* London: J. M. Dent.

Gazzaniga, M. S. (2000). Cerebral specialization and interhemispheric communication: Does the corpus callosum enable the human condition? *Brain, 123,* 1293–1326.

Gazzaniga, M. S. (2005). Forty-five years of split-brain research and still going strong. *Nature Reviews: Neuroscience, 6,* 653–659.

Gazzaniga, M. S. (2015). *Tales from both sides of the brain.* New York: HarperCollins.

Gazzaniga, M. S., Bogen, J. E., & Sperry, R. W. (1962). Some functional effects of sectioning the cerebral commissures in man. *Proceedings of the National Academy of Sciences, 48,* 1765–1769.

Geraldo, C. (2017, November 2). Campuses face mental health crisis. Retrieved from https://www.tmjr.com/news/i-team/campuses-face-mental-health-crisis.

Geschwind, N., & Galaburda, A. M. (1987). *Cerebral lateralization: Biological mechanisms, associations, and pathology.* Cambridge, MA: Bradford Books/MIT press.

Gieryn, T. F. (1999). *Cultural boundaries of science: Credibility on the line.* Chicago: University of Chicago Press.

Gilman, C. P. (1913, October). Why I wrote "The Yellow Wall-Paper." *The Forerunner.* On-line reprint retrieved from https://csivc.csi.cuny.edu/history/files/lavender/whyyw.html.

Gilman, C. P. (1963). *The living of Charlotte Perkins Gilman: An autobiography.* New York: Harper & Row.

Gilman, C. P. (2000). *The yellow wall-paper.* New York: The Modern Library. (reprinted from 1892, *The New England Magazine,* pp. 647–656).

Glover, R. M. (1851). Lectures on the philosophy of medicine. *The Lancet, 1428,* 35–38.

Golinski, J. (1998). *Making natural knowledge: Constructivism and the history of science.* Cambridge: Cambridge University Press.

Gordon, M. (1904). Treatment of inebriety by drugs. *The British Medical Journal, 1*(2266), 1345–1346.

Gordon, S. (2015). Medical condition, demon or undead corpse? Sleep paralysis and the nightmare in medieval Europe. *Social History of Medicine, 28,* 425–444.

Gosling, F. G. (1987). *Before Freud: Neurasthenia and the American Medical Community.* Urbana: University of Illinois Press.

Gottesman, C. (2009). Discovery of the dreaming sleep stage: A recollection. *Sleep, 32,* 15–16.

Graziano, A. B., & Johnson, J. K. (2015). Music, neurology, and psychology in the nineteenth century. *Progress in Brain Research, 216,* 33–49.

Green, A. R. (2008). Gaddum and LSD: The birth and growth of experimental and clinical neuropharmacology research on 5-HT in the UK. *British Journal of Pharmacology, 154,* 1583–1599.

Greene, J. A. (2007). *Prescribing by numbers: Drugs and the definition of disease.* Baltimore: Johns Hopkins University Press.

Greene, R. H., & Brooks, H. (1912). *Diseases of the genito-urinary organs and the kidney.* Philadelphia: W. B. Saunders.

Greenfield, S. (2016). *A day in the life of the brain.* London: Penguin Books.

Gregory, J. (2007). *Of Victorians and vegetarians.* London: Tauris Academic Studies.

Grill, C. (2013, April 1). State law update: Lawmakers address prescribing practices and pill mills. *Bulletin of the American College of Surgeons.* Retrieved February 1, 2014, at http://bulletin.facs.org/2013/04/prescribing-pill-mills/.

Grinspoon, L., & Bakalar, J. B. (1979). *Psychedelic drugs reconsidered.* New York: Basic Books.

Grob, C. S., Bossis, A. P., & Griffiths, R. R. (2013). Use of the classic hallucinogen psilocybin for the treatment of existential distress associated with cancer. In B. I. Carr, & J. Steel (Eds.)., *Psychological aspects of cancer* (pp. 291–308). New York: Springer.

Grosskurth, P. (1980). *Havelock Ellis: A biography.* London: Allen Lane.

Guttmann, E., & Maclay, W. S. (1936). Mescalin and depersonalization: Therapeutic experiments. *The Journal of Neurology and Psychopathology, 16,* 193–212.

Haas, S. V. (1924). The value of the banana in the treatment of celiac disease. *The American Journal of Diseases of Children, 24,* 421–437.

Haas, S. V., & Haas, M. (1950). Diagnosis and treatment of celiac disease. *Postgraduate Medicine, 7,* 239–250.

Hadlock, H. (2000). Sonorous bodies: Women and the glass harmonica. *Journal of the American Musicological Society, 53,* 507–542.

Halberstadt, A. L. (2015). Recent advances in the neuropsychopharmacology of serotonergic hallucinogens. *Behavior and Brain Research, 277,* 99–120.

Hall, G. S. (1904). *Adolescence.* New York: Appleton.

Hall, L. A. (1992). Forbidden by God, despised by men: Masturbation, medical warnings, moral panic, and manhood in Great Britain, 1850–1950. *Journal of the History of Sexuality, 2,* 365–387.

Hall, M. (1837). *Memoirs on the nervous system.* London: Sherwood, Gilbert & Piper.

Hammond, W. A. (1869). *Sleep and its derangements.* Philadelphia: J. B. Lippincott.

Hanslick, E. (1891). *The beautiful in music* (7th ed.). London: Novello, Ewer.

Hare, E. H. (1962). Masturbatory insanity: The history of an idea. *The Journal of Mental Science, 108,* 1–25.

Harrington, A. (1987). *Medicine, mind, and the double brain.* Princeton: Princeton University Press.

Harris, L. J. (1984). Louis Pierre Gratiolet, Paul Broca, et al. on the question of a maturational left-right gradient: Some forerunners of current day models. *Behavioral & Brain Sciences, 7,* 730–731.

Harris, L. J. (1999). Early theory and research on hemispheric specialization. *Schizophrenia Bulletin, 25,* 11–39.

Harrison, J. P. (1825). Observations on Gall and Spurzheim's theory. *Philadelphia Journal of Medical and Physical Sciences, 11,* 233–249.

Hartley, D. (1749). *Observations on man, his frame, his duty, and his expectations.* London: S. Richardson.

Hartley, H., & Tiefer, L. (2003). Taking a biological turn: The push for a "female Viagra" and the medicalization of female sexual problems. *Women's Studies Quarterly, 31*(1/2), 42–54.

Hartley, L. (2001). *Physiognomy and the meaning of expression in nineteenth century culture.* Cambridge: Cambridge University Press.

Haubrich, W. S. (1993). Sylvester Graham: Partly right, mostly for the wrong reasons. *Journal of Medical Biography, 6,* 240–243.

Healy, D. (2012). *Pharmaggedon*. Berkeley: University of California Press.

Hebb, D. O. (1949). *The organization of behavior.* New York: John Wiley & Sons.

Helfand, W. H. (1974). James Morison and his pills. *Transactions of the British Society for the History of Pharmacy, 1,* 101–135.

Helfand, W. H. (1996). Selling addiction cures. *Transactions and Studies of the College of Physicians of Philadelphia, 18,* 85–108.

Helfand, W. H. (2002). *Quack, quack, quack: The sellers of nostrums in prints, posters, ephemera and books.* New York: The Grolier Club.

Helmholtz, H. (1954). *On the sensations of tone.* New York: Dover. (Original work published 1863.)

Herman, L. (1938). *The practice of urology.* Philadelphia: W.B. Saunders.

Hermes, M., Hagemann, D., Naumann, E., & Walter, C. (2011). Extraversion and its positive emotional core—Further evidence from neuroscience. *Emotion, 11,* 367–378.

Hermle, L., Spitzer, M., & Gouzoulis, E. (1994). Arylalkanamine-induced effects in normal volunteers: On the significance of research in hallucinogenic agents for psychiatry. In A. Pletscher, & D. Ladewig (Eds.). *50 years of LSD: Current status and perspectives of hallucinogens* (pp. 87–99). New York: Parthenon.

Herzberg, D. (2009). *Happy pills in America: From Miltown to Prozac.* Baltimore: Johns Hopkins University Press.

Hinman, F. (1935). *The principles and practice of urology*. Philadelphia: W.B. Saunders.

Hirschbein, L. D. (2000). The glandular solution: Sex, masculinity and the Aging in the 1920s. *The Journal of the History of Sexuality, 9,* 277–304.

Hobson, J. A. (2005). Sleep is of the brain, by the brain and for the brain. *Science, 437,* 1254–1256.

Hodgson, B. (1999). *Opium: A portrait of the heavenly demon.* London: Souvenir Press.

Hoffer, A. (1970). Treatment of alcoholism with psychedelic therapy. In B. Aaronson, & H. Osmond (Eds.). *Psychedelics: The uses and implications of hallucinogenic drugs.* New York: Bobbs-Merrill.

Hoffer, A., & Osmond, H. (1967). *The hallucinogens.* New York: Academic Press.

Hofmann, A. (1970). The discovery of LSD and subsequent investigations on naturally occurring hallucinogens. In F. J. Ayd, & B. Blackwell (Eds.), *Discoveries in biological psychiatry* [chapter reprint]. Philadelphia: Lippincott.

Holland, H. (1852). *Chapters on mental physiology.* London: Longman, Brown, Green, & Longmans.

Hopkins, J. E. (1883). *True manliness.* London: Hatchards.

Humphreys, K. (2012). What can we learn from the failure of yet another "miracle cure" for addiction? *Addiction, 107,* 237–239.

Insel, T. (2006, May 9). Mental health research: Into the future. [Director's Report]. National Institute of Mental Health. Retrieved from http://www.nimh.nih.gov/about/updates/2006/mental-health-research-into-the-future.shtml.

Ireland, W. W. (1880). Notes on left-handedness. *Brain, 3,* 207–214.

Ireland, W. (1886). *The blot upon the brain: Studies in history and psychology.* New York: G. P. Putnam's Sons.

Ireland, W. W. (1891). On the discordant action of the double brain. *British Medical Journal, 1,* 1167–1169.

Ishimori, K. (1909). True cause of sleep: A hypnogenic substance as evidenced in the brain of sleep-deprived animals. *Tokyo Igakkai Zasshi, 23,* 429–457.

Jackson, J. H. (1868). Hemispherical coordination. *Medical Times and Gazette, 2,* 208–209, 358–359.

Jacyna, L. S. (1981). The physiology of mind, the unity of nature, and the moral order in Victorian thought. *British Journal for the History of Science, 14,* 109–132.

Jacyna, L. S. (1982). Somatic theories of mind and the interests of medicine in Britain, 1850–1879. *Medical History, 26,* 233–258.

Jacyna, L.S, & Casper, S.T. (2012). Introduction. In L.S. Jacyna, & S.T. Casper (Eds.). *The neurological patient in history* (pp. 1–11). Rochester: University of Rochester Press.

James, W. (1882). On some Hegelisms. *Mind, 7,* 206–208.

James, W. (1890). *Principles of psychology* (2 Vols.). New York: Henry Holt.

James, W. (1898). Consciousness under nitrous oxide. *Psychological Review, 5,* 194–196.

James, W. (1902). *The varieties of religious experience.* New York: Longmans, Green.

James, W. (1978). A pluralistic mystic. In *Essays in philosophy* (pp. 172–190). Cambridge: Harvard University Press. (Original work published 1910.)

Jameson, E. (1961). *The natural history of quackery.* London: Michael Joseph.

Jang, K. L., Livesley, W. J., & Vernon, P. A. (1996). Heritability of the big five personality dimensions and their facets: A twin study. *Journal of Personality, 64,* 577–592.

Jaspers, L., Feys, F., Bramer, W. M., Franco, O. H., Leusink, P., & Laan, E. T. M. (2016). Efficacy and safety of Flibanserin for the treatment of hypoactive sexual desire disorder in women: A systematic review and meta-analysis. *Journal of the American Medical Association: Internal Medicine, 176,* 453–462.

Jay, M. (2010). *High society.* Rochester, VT: Park Street Press.

John, O. P., & Srivastava, S. (1999). The big five trait taxonomy: History, measurement, and theoretical perspectives. In L. A. Pervin, & O. P. John (Eds.), *Handbook of personality: Theory and research* (2nd ed.) (pp. 102–138). New York: Guilford.

Johnson, R. L. (2011). "Disease is unrhythmical": Jazz, health, and disability in 1920s America. *Health and History, 13* (2), 13–42.

Johnston, J., & Parens, E. (Eds.). (2014). *Interpreting neuroimages: An introduction to the technology and its limits.* Hastings Center Report, 45 (2) [special report].

Jones, J. (1700). *Mysteries of opium revealed.* London: Richard Smith.

Jouvet, M. (1965). Paradoxical sleep: A study of its nature and its mechanism. *Progress in Brain Research, 18,* 20–57.

Jouvet, M. (1967). Neurophysiology of the states of sleep. *Physiological Reviews, 47,* 117–177.

Jouvet, M. (1999). *The paradox of sleep: The story of dreaming.* [trans. L. Garey]. Cambridge: MIT Press.

Joyce, K. A. (2008). *Magnetic appeal: MRI and the myth of transparency.* Ithaca: Cornell University Press.

Jung, C. (1971). *Psychological types.* Princeton: Princeton University Press.

Kagan, J. (2013). Equal time for psychological and biological contributions to human variation. *Review of General Psychology, 17,* 351–357.

Kagan, J., & Snidman, N. (2004). *The long shadow of temperament.* Cambridge: Harvard University Press.

Kalant, O. J. (1987). *Maier's cocaine addiction (Der Kokainismus).* Toronto: Addiction Research Foundation [Trans. O. J. Kalant]. (Original worked published 1926, Leipzig: Georg Thieme Verlag.)

Kandall, S. R. (1996). *Substance and shadow: Women and addiction in the U.S.* Cambridge: Harvard University Press.

Karch, S. B. (2006). *A brief history of cocaine.* Boca Raton, FL: CRC Press.

Karmonik, C., Brandt, A., Anderson, J. R., Brooks, F., Lytle, J., Silverman, E., & Frazier, J. T. (2016). Music listening modulates functional connectivity and information flow in the human brain. *Brain Connectivity, 6*(8), 632–641.

Kast, E. (1967). Attenuation and anticipation: A therapeutic use of lysergic acid diethylamide. *Psychiatric Quarterly, 41,* 646–657.

Keeley, L. E. (1892a). "Cures" for inebriety: Letter to the editor. *Lancet, II,* 285–286.

Keeley, L. E. (1892b). Drunkenness, a curable disease. *American Journal of Politics, 1,* 27–43.

Kempf, E. J. (1918). *The autonomic functions and the personality.* Washington, D.C.: Nervous & Mental Disease Publications.

Kennaway, J. (2010). From sensibility to pathology: The origins of the idea of nervous music around 1800. *Journal of the History of Medical and Allied Sciences, 65,* 396–426.

Kennaway, J. (2012). *Bad vibrations.* Farnham, UK: Ashgate.

Kerr, N. (1894). *Inebriety or narcomania: Its etiology, pathology, treatment and jurisprudence* (3rd ed.). London: H. K. Lewis.

Keyes, E. L. (1923). *Urology*. New York: D. Appleton.

Klein, M. (1984). *Envy and gratitude and other works, 1946–1963*. New York: Macmillan. (Original work published 1959.)

Klüver, H. (1966). *Mescal and mechanisms of hallucinations*. [reprint 1928]. Chicago: University of Chicago Press.

Knauer, A., & Maloney, W. J. M. A. (1913). A preliminary note on the psychic action of mescaline, with special reference to the mechanism of visual hallucinations. *Journal of Nervous and Mental Disease, 40*, 425–439.

Koch, C., & Marcus, G. (2014). Cracking the brain's codes. *MIT Technology Review*. [on-line article]. Retrieved from http://www.technologyreview.com/s/528131/cracking-the-brains-codes.

Kochanek, K. D., Murphy, S. L., Xu, J., & Arias, E. (2017, December). Mortality in the United States, 2016. *NCHS Data Brief, No. 293*. Retrieved from https://www.cdc.gov/nchs/ data/databriefs/db293.pdf.

Koelsch, S. (2013). *Brain and music*. Chichester, UK: Wiley-Blackwell.

Kolata, G., & Cohen, S. (2016, January 17). Drug overdoses propel rise in mortality rates of Whites. *The New York Times, 165*, 1, 18.

Kroker, K. (2007). *The sleep of others and the transformations of sleep research*. Toronto: University of Toronto Press.

Kroll, J., & Bachrach, B. (2005). *The mystic mind: The psychology of medieval mystics and ascetics*. New York: Routledge.

Kumar, V. M. (2015). Sleep medicine in ancient and traditional India. In S. Chokroverty, & M. Billiard (Eds.), *Sleep medicine* (pp. 25–28). New York: Springer.

Kurtzweil, P. (1995). Medical possibilities for psychedelic drugs. *FDA Consumer Magazine, 29*(7), 5 pp. [online journal]. Retrieved from www.fda.gov/fdac/features/795_psyche.html.

Labatt, S. B., & Stokes, W. (1833). A case of erysipelas terminating fatally, with symptoms of gastritis, after the use of a large quantity of "Morison's pills." *Dublin Journal of Medical and Chemical Science, 4*, 237–241.

Lashley, K. (1950). In search of the engram. *Symposium for the Society of Experimental Biologists, 4*, 454–482.

Lavater, J. C. (1789). *Essays on physiognomy*. London: G. G. J. & J. Robinson.

Lawlor, C. (2012). *From melancholia to Prozac*. Oxford: Oxford University Press.

Leary, D. (1985). Scientific amnesia. *Behavioral and Brain Sciences, 8*, 641–642.

Lee, M. A., & Shlain, B. (1985). *Acid dreams: The CIA, LSD, and the sixties rebellion*. New York: Grove Weidenfeld.

Legrenzi, P., & Umiltá, C. (2011). *Neuromania: On the limits of brain science* [F. Anderson, Trans.]. New York: Oxford University Press.

Lenoir, T. (1994). Helmholtz and the materialities of communication. *Osiris, 9*, 184–207.

Levenstein, H. A. (1988). *Revolution at the table*. New York: Oxford University Press.

Levinstein, E. (1878). *Morbid craving for morphia* [trans. C. Harrer]. London: Smith, Elder.

Lewin, L. (1998). *Phantastica: A classic survey on the use and abuse of mind-altering plants*. Trans. P. H. A. Wirth [Reprint of 1927 German edition]. Rochester, VT: Park Street Press.

Lieberman, E. J. (1962). Psychochemicals as weapons. *Bulletin of Atomic Scientists, 18*(1), 11–14.

Ling, W., Shoptaw, S., Hillhouse, M., Bholat, M. A., Charuvastra, C., Heinzerling, K., ..., & Doraimani, G. (2011). Double-blind placebo-controlled evaluation of the PROMETA™ protocol for methamphetamine dependence. *Addiction, 107*, 361–369.

Littlefield, M., & Johnson, J. M. (Eds.). (2012). *The neuroscientific turn: Transdisciplinarity in the age of the Brain*. Ann Arbor: University of Michigan Press.

Loe, M. (2004). *The rise of Viagra*. New York: New York University Press.

Logan, P. M. (1997). *Nerves and narratives: A cultural history of hysteria in 19th century British prose*. Berkeley: University of California Press.

Lokhorst, G-J. (1982). An ancient Greek theory of hemispheric specialization. *Clio Medica, 17*, 33–38.

Lombroso, C. (1903). Left-handedness and left-sidedness. *The North American Review, 177*, 440–444.

Lowsley, O. S., & Kirwin, T. J. (1926). *A textbook of urology*. Philadelphia: Lea & Febiger.

Loza, S. (2015). Sleep medicine in the Arab Islamic civilization. In S. Chokroverty, & M. Billiard (Eds.), *Sleep medicine* (pp. 21–24). New York: Springer.

Luck, S. J., Hillyard, S. A., Mangun, G. R., & Gazzaniga, M. S. (1989). Independent hemispheric attentional systems mediate visual search in split-brain patients. *Nature, 342,* 543–545.

Lyons, S. L. (2009). *Species, serpents, spirits, and skulls*. Albany, NY: SUNY Press.

MacNish, R. (1842). *The philosophy of sleep* (2nd ed.). Hartford, CT: William Andrews.

Maher, B. A., & Maher, W. B. (1994). Personality and psychopathology: A historical perspective. *Journal of Abnormal Psychology, 103,* 72–77.

Maisto, S. A., Galizio, M., & Connors, G. J. (2015). *Drug use and abuse* (7th ed.). Stamford, CT: Cengage Learning.

Margulis, E. H. (2008). Neuroscience, the food of musical culture? *Review of General Psychology, 12,* 159–169.

Marshall, B. L. (2002). "Hard science": Gendered constructions of sexual dysfunction in the "Viagra Age." *Sexualities, 5*(2), 131–158.

Marshall, B. L., & Katz, S. (2002). Forever functional: Sexual fitness and the aging male body. *Body & Society, 8*(4), 43–70.

Marshall, V. F. (1956). *Textbook of urology*. New York: Hoeber-Harper.

Masson, J. M. (Ed.). (1985). *The complete letters of Sigmund Freud to Wilhelm Fliess 1887–1904*. [trans. J. M. Masson]. Cambridge: Harvard University Press.

Matfin, G. (2010). The rejuvenation of testosterone: Philosopher's stone or Brown-Sèquard's Elixir? *Therapeutic Advances in Endocrinology and Metabolism, 1*(4), 51–54.

Mattison, J. B. (1887). Cocaine dosage and cocaine addiction. *Lancet, I,* 1024–1026.

Maudsley, H. (1889). The double brain. *Mind, 14,* 161–187.

Mayer, E. (2016). *The mind-gut connection*. New York: HarperCollins.

McCabe, S. (2000). Rapid detox: Understanding new treatment approaches for the addicted patient. *Perspectives in Psychiatric Care, 36,* 113–119.

McCrae, L. E. (1948). *Clinical urology* (2nd ed.). Philadelphia: F. A. Davis.

McCrae, R. R., & Costa, P. T. (1987). Validation of the five-factor model of personality across instruments and observers. *Journal of Personality and Social Psychology, 52,* 81–90.

McGuire, W. (Ed.). (1974). *The Freud/Jung letters*. R. Manheim, & R. F. C. Hull, Trans. Princeton: Princeton University Press.

Merlo, L.J., Singhakant, S., Cummings, S.M., & Cottler, L. B. (2013). Reasons for misuse of prescription medication among physicians undergoing monitoring by a physician health program. *Journal of Addiction Medicine, 7*(5), 349 DOI: 10.1097/ADM.0b013e31829da074.

Milligan, B. (2005). Morphine-addicted doctors, the English opium-eater, and embattled medical authority. *Victorian Literature and Culture, 33,* 541–553.

Mischel, W. (2015). *The marshmallow test: Why self-control is the engine of success*. New York: Little, Brown.

Mitchell, S. W. (1872). *Wear and Tear, or Hints for the Overworked*. Philadelphia: Lippincott.

Mitchell, S. W. (1885). *Lectures on diseases of the nervous system especially in women* (2nd ed.). Philadelphia: Lea Brothers.

Mitchell, S. W. (1888). *Doctor and patient* (2nd ed.). Edinburgh: Young J. Pentland.

Mitchell, S. W. (1896, December 5). Remarks on the effects of *Anhelonium Lewinii* (the mescal button). *British Medical Journal, 2,* 1625–1629.

Mitchell, S. W. (1898). *Fat and blood and how to make them* (7th Ed.). Philadelphia: Lippincott.

Mitchell, S. W. (1904). The evolution of the rest treatment. *Journal of Nervous and Mental Diseases* [reprint], 1–6.

Moat, T. (1832). *The practical proofs of the soundness of the Hygeian system of physiology.* Brooklyn, NY: British College of Health.

Mooney, J. (1896). Mescal plant and ceremony. *Therapeutic Gazette, 20,* 7–11.

Moreno, F. A., Wiegand, C. B., Taitano, E. K., & Delgado, P. L. (2006). Safety, tolerability, and efficacy of psilocybin in 9 patients with obsessive-compulsive disorder. *Journal of Clinical Psychiatry, 67,* 1735–1740.

Morison's Hygeian pills. (1836). *The Boston & Surgical Journal, 14,* 206–207.

Morton, H. H. (1919). *Genitourinary diseases and syphilis* (4th ed.). London: Henry Kimpton.

Mumford, K. J. (1992). "Lost manhood" found: Male sexual impotence and Victorian culture in the U. S. *Journal of the History of Sexuality, 3,* 33–57.

Musto, D. F. (1999). *The American disease: Origins of narcotic control* (3rd ed.). New York: Oxford University Press.

National Institute of Drug Abuse (NIDA). (2016). National overdose deaths. Retrieved December 1, 2017 from https://www.drugabuse.gov/related-topics/trends-statistics/overdose-death-rates.

Natural language—Phrenology and the new systems of philosophy. (1843). *The Boston Medical and Surgical Journal, 29–30,* 422–423.

Nelson, K. L. (2015a, November 20). Drug overdose death rate for Tennessee youth has doubled over past decade. *Knoxville News Sentinel.* Retrieved from http://search.knoxnews.com/jmg.aspx?k=teen+heroin+overdose.

Nelson, K. L. (2015b, December 11). Tennessee teen drug abuse declines. *Knoxville News Sentinel.* Retrieved from http://www.knoxnews.com/news/local/tenn-teen-drug-abuse-declines-ep-1410728729-361829161.html.

Nestor, J. (2016). Get clean or die trying. *Scientific American, 315*(5), 62–69.

Nolen-Hoeksema, S., & Davis, C. G. (1999). "Thanks for sharing that": Ruminators and their social support networks. *Journal of Personality and Social Psychology, 77,* 801–814.

Novak, S. J. (1997). LSD before Leary: Sidney Cohen's critique of 1950s psychedelic drug research. *Isis, 88,* 87–110.

O'Donnell, J. K., Halpin, J., Mattson, C. L., Goldberger, B. A., & Gladden, R. M. (2017). Deaths involving fentanyl, fentanyl analogs, and U-47700-10 states, July–December 2016. *CDC Morbidity & Mortality Weekly Report, 66*(43), 1197–1202.

O'Kelly, J., Fachner, J. C., & Tervaniemi, M. (2016, November 22). Editorial. Dialogues in music therapy and music neuroscience: Collaborative understanding driving clinical advances. *Frontiers in Human Neuroscience, 10,* Article 585 [On-line publication]. Doi: 10.3389/fnhum.2016.00585.

Oppenheim, J. (1991). *Shattered nerves: Doctors, patients, and depression in Victorian England.* Oxford: Oxford University Press.

Oram, M. (2014). Efficacy and enlightenment: LSD psychotherapy and the drug amendments of 1962. *Journal of the History of Medical and Allied Sciences, 69,* 250–280.

O'Regan, E. (2017, November 3). We live in an age of anxiety: Surge in mental health problems in last six years. *The Irish Independent.* Retrieved from https://www.independent.ie/irish-news/health/we-live-in-an-age-of-anxiety-surge-in-mental-health-problems-in-last-six-years-36285433.html.

Osler, W. (1901). *The principles and practice of medicine* (4th ed.). Edinburgh: Young J. Pentland.

Osmond, H., & Smythies, J. (1952). Schizophrenia: A new approach. *British Journal of Psychiatry, 98,* 309–315.

Ostojic, S. M. (2006). Yohimbine: The effects on body composition and exercise performance in soccer players. *Research in Sports Medicine, 14,* 289–299.

Parssinen, T. (1974). Popular science and society: The phrenology movement in early Victorian Britain. *Journal of Social History, 8*(1), 1–20.

Parssinen, T. (1983). *Secret passions, secret remedies: Narcotic drugs in British society 1820–1930.* Philadelphia: Institute for the Study of Human Issues.

Parssinen, T.M., & Kerner, K. (1980). Development of the disease model of drug addiction in Britain, 1870–1926. *Medical History, 24,* 275–296.

Parton, I. (1960). *Urology in general practice.* London: Butterworth.

Passie, T. (1997). *Psycholytic and psychedelic therapy research 1931–1995. A complete international bibliography.* Hanover, Germany: Laurentius.

Perry, R. B. (1948). *The thought and char-*

acter of William James. Nashville: Vanderbilt University Press.
Peters, D. (1981). The British medical response to opiate addiction in the nineteenth century. *Journal of the History of Medicine and Allied Sciences, 36*, 455–488.
Peterson, M. J. (1986). Dr. Acton's enemy: Medicine, Sex and Society in Victorian England. *Victorian Studies, 29*, 569–590.
Phrenology. (1824). *The Lancet, 1*, 495–497.
Piéron, H. (1913). *Le problem physiologique du sommeil*. Paris: Masson.
Pinto, Y., Neville, D. A., Otten, M., Corballis, P. M., Lamme, V. A. F., de Haan, E. H. F., ..., & Fabri, M. (2017). Split brain: Divided perception but undivided consciousness. *Brain* [on-line article, 7 pp.]. doi: 10.1093/brain/aww358.
Pletscher, A., & Ladewig, D. (Eds.). (1994). *50 years of LSD: Current status and perspectives of hallucinogens*. New York: Parthenon.
Poirier, S. (1983). The Weir Mitchell rest cure: Doctor and patients. *Women's Studies, 10*, 15–40.
Portenoy, R., & Foley, K. (1986). Chronic use of opioid analgesics in non-malignant pain: Report of 38 cases. *Pain, 25*, 171–186.
Porter, J., & Jick, H. (1980). Addiction rare in patients treated with narcotics. *New England Journal of Medicine, 302*(2), 123.
Porter, R. (1989). *Health for sale*. Manchester: Manchester University Press.
Porter, R. (2000). *Quacks*. Charleston, SC: Tempus.
Porter, R. (2003a). *Flesh in the age of reason*. New York: W. W. Norton.
Porter, R. (2003b). *Quacks: Fakers, charlatans in medicine*. Stroud, UK: Tempus.
Prentiss, D. W., & Morgan, T. P. (1895). *Anhelonium lewinie* (mescal buttons): A study of the drug with especial reference to its physiological action upon man, with reports of experiments. *Therapeutic Gazette, 9*, 577–585.
Puente, A. (1993). Sperry's emergent interactionism as a new explanation for dualism. *Journal of General Psychology, 120*, 65–72.
Quinones, S. (2015). *Dreamland*. New York: Bloomsbury Press.
Raby, W. N. (2000). Gabapentin therapy for cocaine cravings. *American Journal of Psychiatry, 157*, 2058–2059.
Rather, L. S. (1968). Six things non-natural. *Clio Medica, 3*, 337–347.
Rauscher, F. H., Shaw, G. L., & Ky, K. N. (1993). Music and spatial task performance. *Nature, 365*, 611.
Rayner, J. (2017). *The angry chef: Bad science and the truth about healthy eating*. London: One World.
Rea, K., Dinan, T. G., Cryan, J. F. (2016). The microbiome: A key regulator of stress and neuroinflammation. *Neurobiology of Stress, 4*, 23–33.
Reiser, S. J. (1993). The science of diagnosis: Diagnostic technology. In W. B. Bynum, & R. Porter (Eds.). *The companion* encyclopedia *of the history of medicine* (Vol. II, pp. 826–851). London: Routledge.
Reiser, S. J. (2009). *Technological medicine*. Cambridge: Cambridge University Press.
Report of the British Phrenological Society. (1842). *The Lancet, II*, 703.
Report of the Society for the Study of Inebriety. (1892). *The Lancet, II*, 106–107.
Review of J. Spurzheim's *Anatomy of the Brain*. (1826). *The Lancet, I*, 200–211.
Richter, R. (1994). Psychotherapeutic effects. In A. Pletscher, & D. Ladewig (Eds.). *50 years of LSD: Current status and perspectives of hallucinogens* (pp. 203–211). New York: Parthenon.
Ridgway, E. S. (1993). John Elliotson (1792–1868): A bitter enemy of legitimate medicine? Part I: Earlier years and the introduction of mesmerism. *The Journal of Medical Biography, 1*, 191–198.
Robinson, D. L. (2001). How brain arousal systems determine different temperament types and the major dimensions of personality. *Personality and Individual Differences, 31*, 1233–1259.
Robinson, T.E., & Berridge, K.C. (1993). The neural basis of drug craving: An incentive-sensitization theory of addiction. *Brain Research Reviews, 18*, 247–291.
Robinson, W. J. (1912). *A practical treatise on the causes, symptoms, and treatment of sexual impotence*. New York: Eugenics.

Rogers, C. R. (1959). A theory of therapy, personality, and interpersonal relationships, as developed in the client-centered framework. In S. Koch (Ed.), *Psychology: A study of a science* (Vol. 3). New York: McGraw-Hill.

Rose, N., & Abi-Rached, J.M. (2013). *Neuro: The new brain sciences and the management of the mind.* Princeton: Princeton University Press.

Rosenberg, C. E. (1997). *No other gods: On science and American social thought.* Baltimore: Johns Hopkins University Press.

Rothbart, M. K., Derryberry, D., & Posner, M. I. (1994). A psychobiological approach to the development of temperament. In J. E Bates, & T. D. Wachs (Eds.), *Temperament: Individual differences and the interface of biology and behavior* (pp. 83–116). Washington, D.C.: American Psychological Association.

Rotunda, M. (2007). Savages to the left of me, neurasthenics to the right, stuck in the middle with you: Inebriety and human nature in American Society, 1855–1900. *Canadian Bulletin of Medical History, 24,* 49–65.

Rousseau, G. (2007). Temperament and the long shadow of nerves in the eighteenth century. In H. Whitaker, C. U. M. Smith, & S. Finger (Eds.), *Brain, mind and medicine: Essays in eighteenth century neuroscience* (pp. 353–369). New York: Springer.

Rousseau, G. (2014). (Nervously) grappling with (musical) "pictures in the mind": A personal account. In J. Kennaway (Ed.), *Music and the nerves, 1700–1900* (pp. 18–43). Basingstoke, UK: Palgrave Macmillan.

Rudolf, R. D. (1922). The treatment of insomnia. *British Medical Journal, I,* 377–379.

Rushton, R., & Steinberg, H. (1963). Mutual potentiation of amphetamine and amylobarbitone measured by activity in rats. *British Journal of Pharmacology, 21,* 295–305.

Rylance, R. (2000). *Victorian psychology and British culture 1850–1880.* New York: Oxford University Press.

Sacks, O. (2012). *Hallucinations.* New York: Knopf.

Santhouse, A. M., Howard, R. J., & ffytche, D. H. (2000). Visual hallucinatory syndromes and the anatomy of the visual brain. *Brain, 123,* 2055–2064.

Saper, C. B., Scammell, T. E., & Lu, J. (2005). Hypothalamic regulation of sleep and circadian rhythms. *Nature, 437,* 1257–1263.

Satel, S., & Lilienfeld, S.O. (2013). *Brainwashed: The seductive appeal of mindless neuroscience.* New York: Basic Books.

Schellenberg, E. G. (2013). Cognitive performance after listening to music: A review of the Mozart effect. In R. MacDonald, G. Kreutz, & L. Mitchell (Eds.), *Music, health, and well-being* (pp. 324–338). Oxford: Oxford University Press.

Schulte, B. (2015, August 18). From 1952 to 2015: The path to "female Viagra" has been a rocky one. *The Washington Post.* Retrieved from https://www.washingtonpost.com/news/to-your-health/wp/2015/08/17/female-viagra-could-get-fda-approval-this-week/?utm_term=.90115197fc84.

Schultes, R. E. (1938). The appeal of peyote (*lophophora williamsii*) as a medicine. *American Anthropologist, 40,* 698–715.

Schulz, H., & Salzarulo, P. (2015). The evolution of sleep medicine in the nineteenth and the early 20th century. In S. Chokroverty & M. Billiard (Eds.), *Sleep medicine* (pp. 75–89). New York: Springer.

Schuster, D. (2006). *Neurasthenic nation.* New Brunswick: Rutgers University Press.

Schwartz, L. M., & Woloshin, S. (2009). Lost in transmission—FDA drug information that never reaches clinicians. *The New England Journal of Medicine, 361,* 1717.

Scott, D. (1977). Musicogenic epilepsy. In M. Critchley, & R. A. Henson (Eds.), *Music and the brain* (pp. 354–364). London: Wm. Heinemann Medical Books.

Selye, H. (1978). *The stress of life* (2nd ed.). New York: McGraw-Hill.

Severance, E. G., Prandovszky, E., Castiglione, J., & Yolken, R. H. (2015). Gastroenterology issues in schizophrenia: Why the gut matters. [on-line article]. *Current Psychiatry Reports, 17*(5), 27pp.

Severance, E. G., Yolken, R. H., & Eaton, W. W. (2016). Autoimmune diseases, gastrointestinal disorders and the microbiome in schizophrenia: More than a gut feeling. *Schizophrenia Research, 176*(1), 23–35.

Sewall, T. (1839). *An examination of phrenology.* Boston: D. S. King.

Shapin, S. (2003). Trusting George Cheyne: Scientific expertise, common sense, and moral authority in early eighteenth-century dietetic medicine. *Bulletin of the History of Medicine, 77,* 263–297.

Sherrington, C. S. (1906). *The integrative action of the nervous system.* New Haven: Yale University Press.

Shimada-Sugimoto M., Otowa T., & Hettema, J. M. (2015). Genetics of anxiety disorders: Genetic epidemiological and molecular studies in humans. *Psychiatry and Clinical Neuroscience, 69,* 388–401.

Shortt, S. E. D. (1986). *Victorian lunacy: Richard M. Bucke and the practice of late nineteenth-century psychiatry.* New York: Cambridge University Press.

Shryock, R. H. (1931). Sylvester Graham and the popular health movement. *The Mississippi Valley Historical Review, 18* (2), 172–183.

Sicherman, B. (1977). The uses of a diagnosis: Doctors, patients, and neurasthenia. *Journal of the History of Medicine, 32,* 33–54.

Simon, L. (1998). *Genuine reality: A life of William James.* New York: Harcourt Brace.

Skinner, B. F. (1964). Behaviorism at fifty. In T. W. Wann (Ed.), *Behaviorism and phenomenology* (pp. 79–108). Chicago: University of Chicago Press.

Skrupskelis, I. K., & Berkeley, E. M. (Eds.). (1992a). *The correspondence of William James (Vol. 1): William and Henry 1861–1884).* Chicago: University of Chicago Press.

Skrupskelis, I. K., & Berkeley, E. M. (Eds.). (1992b). *The correspondence of William James (Vol. 2): William and Henry 1885–1896).* Charlottesville: University Press of Virginia.

Smart, R. G., & Storm, T. (1964). The efficacy of LSD in the treatment of alcoholism. *Quarterly Journal on the Study of Alcohol, 25,* 333–338.

Smillie, L. D., & Wacker, J. (2014). Dopaminergic foundations of personality and individual differences. *Frontiers in Human Neuroscience, 8* (Article 874). [on-line article]. DOI: 10.3389/fnhum.2014.00874.

Smith, R. (1992). *Inhibition: History and meaning in the sciences of mind and brain.* Berkeley: University of California Press.

Smith, R. (1993). The background of physiological psychology in natural philosophy. *History of Science, 11,* 75–123.

Smith, R. (1997). *The Norton history of the human sciences.* New York: W. W. Norton.

Snelders, S., Kaplan, C., & Pieters, T. (2006). On cannabis, chloral hydrate, and career cycles of psychotropic drugs in medicine. *Bulletin on the History of Medicine, 80,* 95–114.

Sperry, R. W. (1995). The future of psychology. *American Psychologist, 50,* 505–506.

Spurzheim, J. (1815). *Outlines of the physiognomical system of Drs. Gall and Spurzheim.* London: Baldwin, Cradock & Joy.

Spurzheim, J. (1833). *Phrenology, in connexion with the study of physiognomy.* Boston: Marsh, Capon & Lyon.

Squire, A. O. (1892). Domestic correspondence. *The Journal of the American Medical Association, 18,* 591–592.

Stall, S. (1897). *What a young boy ought to know.* Philadelphia: Vir.

Stern, J. (2015). Musicogenic epilepsy. *Handbook of Clinical Neurology, 129,* 469–477.

Stern, M. B. (1971). *Heads and headlines: The phrenological Fowlers.* Norman: University of Oklahoma Press.

Stetson, C. P. (1898). The yellow wall-paper. *The New England Magazine.* Reprinted in J. B. Dock (Ed.). (1998). *Charlotte Perkins Gilman's "The yellow wall-paper" and the history of its publication and reception: A critical edition and documentary casebook* (pp. 27–42). University Park: Pennsylvania State University Press.

Stevens, J. (1987). *Storming heaven: LSD and the American dream.* New York: Grove Press.

Stevens, M. S. (2015). Normal sleep, sleep physiology, and sleep deprivation. *Medscape.* Retrieved from http://emedicine.

medscape.com/article/1188226-overview#a2.

Stewart, J. (1905). Inebriety and the so-called cures. *Bristol Medico-Chirurgical Journal, 23*(88), 144–147.

Stiles, A. (2013). The rest cure, 1873–1925. In D. F. Felluga (ed.), *BRANCH: Britain, Representation and Nineteenth-Century History.* Extension of Romanticism and Victorianism on the Net. Retrieved from http://www.branchcollective.org/?ps_articles=anne-stiles-the-rest-cure-1873–1925.

Stolberg, M. (2000). Self-pollution, moral reform, and the venereal trade: Notes on the sources and historical context of *Onania* (1716). *Journal of the History of Sexuality, 9,* 37–61.

Straus, M. M., Ghitza, U. E., & Tai, B. (2013). Preventing deaths from rising opioid overdose in the U.S.—The promise of naloxone antidote in community-based naloxone take-home programs. *Substance Abuse and Rehabilitation, 4,* 65–72.

Stumpf, C. (2012). *The origins of music.* [Trans. & Ed. D. Trippett]. Oxford: Oxford University Press. (Original work published 1911.)

Substance Abuse and Mental Health Services Administration [SAMSA]. (2014). Prescription drug abuse in the workplace. Retrieved from http://workplace.samhsa.gov/pdf /prescription%20drug%20abuse%20fact%20sheet.pdf.

Swaminathan, N. (2008, June 9). Musicophobia: When your favorite song gives you seizures. *Scientific American* [Online article]. Retrieved from https://www.scientificamerican.com/article/musicophobia-when-your-fa/.

Sykes, I. (2014). Le corps sonore: Music and the auditory body in France 1780–1830. In J. Kennaway (Ed.), *Music and the nerves, 1700–1900* (pp. 72–97). Basingstoke, UK: Palgrave Macmillan.

Symonds, C. P. (1925). Sleep and sleeplessness. *British Medical Journal, I,* 869–871.

Tanner, J. M. (1947). The morphological level of personality. *Proceedings of the Royal Society of Medicine, 40,* 301–305.

Taylor, E. (1996). *William James on consciousness beyond the margin.* Princeton: Princeton University Press.

Taylor, J. M. (1929). Personal glimpses of S. Weir Mitchell. *Annals of Medical History (New Series), 1,* 583–598.

Taylor, K. (2012). *The brain supremacy: Notes from the frontiers of neuroscience.* New York: Oxford University Press.

Thornton, E. M. (1986). *The Freudian fallacy: Freud and cocaine.* London: Paladin.

Thorpy, M. (2011). History of sleep medicine. In P. Montagna, & S. Chokroverty (Eds.), *Handbook of clinical neurology* (Vol. 98, Sleep Disorders, Pt. I) (pp. 3–25). Amsterdam: Elsevier.

Thorpy, M. (2015). Sleep in the seventeenth and eighteenth centuries. In S. Chokroverty, & M. Billiard (Eds.). *Sleep medicine* (pp. 69–72). New York: Springer.

Tissot, S. A. (1832). *Onanism.* New York: Collins & Hannay.

Tomlinson, S. (2005). *Head masters.* Tuscaloosa: University of Alabama Press.

Tone, A. (2009). *The age of anxiety.* New York: Basic Books.

Tononi, G., & Koch, C. (2008). The neural correlates of consciousness: An update. *Annals of the New York Academy of Sciences, 1124,* 239–261.

Tracy, S. W. (2005). *Alcoholism in America: From reconstruction to prohibition.* Baltimore: Johns Hopkins University Press.

Trotter, T. (1807). *A view of the nervous temperament.* London: Longman, Hurst, Rees, & Orme.

Tupper, K. W., Wood, E., Yensen, R., & Johnson, M. W. (2015). Psychedelic medicine: A re-emerging therapeutic paradigm. *Canadian Medical Association Journal, 187,* 1054–1059.

Umanath, S., Sarezky, D., & Finger, S. (2011). Sleepwalking through history: Medicine, arts and courts of law. *Journal of History of the Neurosciences, 20,* 253–276.

Usher, J. E. (1892). "Cures" for inebriety: Letter to the editor. *The Lancet, II,* 345–346.

Uttal, W. R. (2001). *The new phrenology: The limits of localizing cognitive processes in the brain.* Cambridge: MIT Press.

Van Wyck Good, W. (1993). Philadelphia's "literary physician": The papers of Silas

Weir Mitchell. *Transactions and Studies of the College of Physicians of Philadelphia, 15,* 47–56.

Van Wyhe, J. (2004a). *Phrenology and the origins of Victorian scientific naturalism.* Aldershot, UK: Ashgate.

Van Wyhe, J. (2004b). Preface. In J. van Wyhe (Ed.). *Combe's Constitution of Man and nineteenth-century responses* (3 Vols.) (pp. v–xix). Bristol, UK: Thoemmes Continuum.

Vertinsky, B. (2007). Physiques as destiny: William H. Sheldon, Barbara Honeyman Heath and the struggle for hegemony in the science of somatotyping. *Canadian Bulletin of Medical History, 24,* 299–316.

Vollenweider, F. X., Leenders, K. L., Scharfetter, C., Maguire, P., Stadelmann, O., & Angst, J. (1997). Positron emission tomography and flurodeoxyglucose studies of metabolic hyperfrontality and psychopathology in the psilocybin model of psychosis. *Neuropsychopharmacology, 16,* 357–372.

Wager, T. D., Atlas, L. Y., Lindquist, M. A., Roy, M., Woo, C-W., & Kross, E. (2013). An fMRI-based neurologic signature of physical pain. *The New England Journal of Medicine, 368,* 1388–1397.

Wager, T. D., Hernandez, L., & Lindquist, M. A. (2009). Essentials of functional neuroimaging. In G. G. Berntsen, & J. T. Cacioppo (Eds.). *Handbook of neuroscience for the behavioral sciences* (pp. 152–200). Hoboken, NJ: John Wiley & Sons.

Wager, T. D., & Lindquist, M. A. (2011). Essentials of functional magnetic resonance imaging. In J. Decety, & J. T. Cacioppo (Eds.). *The Oxford handbook of social neuroscience* (pp. 69–96). New York: Oxford University Press.

Wakley, T. (1836a). Editorial. *Lancet, I,* 764–765.

Wakley, T. (1836b). Editorial. *Lancet, II,* 57–60.

Wakley, T. (1836c). Editorial. *Lancet, II,* 92–95.

Wakley, T. (1836d). Editorial. *Lancet, II,* 121–124.

Wallace, A. (1899). *The wonderful century.* New York: Dodd, Mead.

Walsh, A. (1974). *Johann Christoph Spurzheim and the rise and fall of scientific phrenology in Boston, 1832–1842.* (Doctoral dissertation). University of New Hampshire, Durham, NH.

Walter, R. D. (1970). *Silas Weir Mitchell, M.D., neurologist: A medical biography.* Springfield, IL: Charles Thomas.

Watson, H. (1836). What is the use of the double brain? *Phrenological Journal and Miscellany, 9,* 608–611.

Watson, J. B. (1913). Psychology as a behaviorist views it. *Psychological Review, 20,* 158–177.

Watson, N. V., & Breedlove, S. M. (2016). *The mind's machine: Foundations of brain and behavior* (2nd ed.). Sunderland, MA: Sinauer.

Weisberg, D. S., Keil, F. C., Goodstein, J., Rawson, E., & Gray, J. R. (2008). The seductive allure of neuroscience explanations. *Journal of Cognitive Neuroscience, 20,* 470–477.

White, W. L. (1998). *Slaying the dragon: The Observations on the nature, cause, and cure of those disorders which are commonly called nervous, hypochondriac or hysteric history of addiction treatment and recovery in America.* Bloomington, IL: Chestnut Health Systems/Lighthouse Institute.

White, W. L. (2008). Alcohol, tobacco and other drug use by addictions professionals: Historical reflections and suggested guidelines. *Alcoholism Treatment Quarterly, 26,* 500–535.

Whytt, R. (1768). In *The works of Robert Whytt* (pp. 487–745). Edinburgh: Balfour, Auld & Smellie.

Wigan, A. L. (1844a). *A new view of insanity: The duality of the mind proved by the structures, functions and diseases of the brain and by the phenomena of mental derangement and shown to be essential to moral responsibility.* London: Longman, Brown, Green, & Longmans.

Wigan, A. L. (1844b). Duality of the mind, proved by the structure, functions, and diseases of the brain. the brain. *Lancet, I,* 39–41.

Williams, S. J. (2005). *Sleep and society.* London: Routledge.

Willis, T. (1692). *The London practice of physick.* London: T. Basset, T. Dring, C. Harper, & W. Crook.

Winslow, F. (1849). The unpublished MSS. of the late Alfred Wigan, M.D., author of the "Duality of Mind," etc. *The Journal of Psychological Medicine and Mental Pathology, 2,* 497–512.

Witt, P. (1954, December). Drugged spiders and webs. *Scientific American,* 79–86.

Woodward, S. B. (1850). Observations on the medical treatment of insanity. *The American Journal of Insanity, 7,* 1–29.

Workman, J. (1863). On certain abdominal lesions in the insane. *The American Journal of Insanity, 20,* 44–60.

Yanjiao, L., Yuping, W., Fang, W., Xue, Y, Yue, H., & Shasha, L. (2015). Sleep medicine in ancient and traditional China. In S. Chokroverty, & M. Billiard (Eds.), *Sleep medicine* (pp. 29–33). New York: Springer.

Yarandi, S. S., Peterson, D. A., Treisman, G. J., Moran, T. H., & Pasricha, P. J. (2016). Modulatory effects of gut microbiota on the central nervous system: How gut could play a role in neuropsychiatric health and diseases. *Journal of Neurogastroenterology and Motility, 22,* 201–212.

Young, H. H., & Davis, D. M. (1926). *Young's practice of urology* (2 Vols.). Philadelphia: W. B. Saunders.

Young, J. H. (1992). *American health quackery.* Princeton: Princeton University Press.

Young, R. M. (1968). The functions of the brain: Gall to Ferrier. *Isis, 59,* 251–268.

Zatorre, R. J., & McGill, J. (2005). Music, the food of neuroscience? *Nature, 434,* 312–315.

Zatorre, R. J., Evans, A., & Meyer, E. (1994). Neural mechanisms underlying melodic perception and memory for pitch. *Journal of Neuroscience, 14,* 1908–1919.

Zeidler, D. L., & Sadler, T. D. (2000, October). Bad science and its social implications: Historical perspectives. Paper presented at the Annual Meeting of the Southeastern Association for the Education of Teachers in Science, Auburn, AL.

Zieger, S. (2008). Victorian hallucinogens. *Romanticism and Victorianism on the Net, 49* [on-line article]. Retrieved from https://www.erudit.org/revue/ravon/2008/v/n49/017857ar.html. DOI : 10.7202/017857ar.

Zisapel, N. (2012). Drugs for insomnia. *Expert Opinion on Emerging Drugs, 17,* 299–317.

Index